Studies in Computational Intelligence

Volume 1025

Series Editor

Janusz Kacprzyk, Polish Academy of Sciences, Warsaw, Poland

The series "Studies in Computational Intelligence" (SCI) publishes new developments and advances in the various areas of computational intelligence—quickly and with a high quality. The intent is to cover the theory, applications, and design methods of computational intelligence, as embedded in the fields of engineering, computer science, physics and life sciences, as well as the methodologies behind them. The series contains monographs, lecture notes and edited volumes in computational intelligence spanning the areas of neural networks, connectionist systems, genetic algorithms, evolutionary computation, artificial intelligence, cellular automata, self-organizing systems, soft computing, fuzzy systems, and hybrid intelligent systems. Of particular value to both the contributors and the readership are the short publication timeframe and the world-wide distribution, which enable both wide and rapid dissemination of research output.

Indexed by SCOPUS, DBLP, WTI Frankfurt eG, zbMATH, SCImago.

All books published in the series are submitted for consideration in Web of Science.

More information about this series at https://link.springer.com/bookseries/7092

Mohiuddin Ahmed · Sheikh Rabiul Islam ·
Adnan Anwar · Nour Moustafa ·
Al-Sakib Khan Pathan
Editors

Explainable Artificial Intelligence for Cyber Security

Next Generation Artificial Intelligence

Springer

Editors
Mohiuddin Ahmed
School of Science
Edith Cowan University
Joondalup, WA, Australia

Sheikh Rabiul Islam
Department of Computing Sciences
University of Hartford
West Hartford, CT, USA

Adnan Anwar
School of IT
Deakin University
Melbourne, VIC, Australia

Nour Moustafa
School of Engineering and IT
UNSW Canberra
Campbell, ACT, Australia

Al-Sakib Khan Pathan
Department of Computer Science
and Engineering
United International University (UIU)
Dhaka, Bangladesh

ISSN 1860-949X   ISSN 1860-9503   (electronic)
Studies in Computational Intelligence
ISBN 978-3-030-96632-4   ISBN 978-3-030-96630-0   (eBook)
https://doi.org/10.1007/978-3-030-96630-0

This Springer imprint is published by the registered company Springer Nature Switzerland AG
The registered company address is: Gewerbestrasse 11, 6330 Cham, Switzerland

Dedicated to

My Loving Son: Zaif Rayan

—Mohiuddin Ahmed

My studious daughter: Farisha Islam

—Sheikh Rabiul Islam

My devoted family

—Adnan Anwar

My family

—Nour Moustafa

My family

—Al-Sakib Khan Pathan

Preface

Cyber security is a very complex and diverse discipline. Numerous technological problems need to be solved to make the world safer. It is evident that there is no sign of a decrease in cyber-crime; instead, it is the opposite in nature due to the unprecedented technological advancement and our reliance on it. The cyber security community has been leveraging artificial intelligence (AI) technology to solve several complex computing problems, e.g., intrusion detection systems to identify malicious network activities. In the past two decades, there have been hundreds of algorithms developed capitalizing on the effectiveness of artificial intelligence. Therefore, we have observed the transition from classical artificial intelligence to deep learning, federated learning, reinforcement learning, etc. These techniques have been critical in providing solutions for cyber security problems. However, most recent variants of artificial intelligence-based methods are being treated as *black-box ones*. There is a lack of explanation that humans can easily understand the solution(s) offered. For example, a particular neural network that is perfect for identifying phishing attacks (i.e., the deception using email) is still obscure due to its complex internal working mechanism. Hence, it is important to explore various avenues of explainable artificial intelligence (XAI), an emerging area of artificial intelligence, to provide a human-friendly decision for cyber security from a broader perspective.

In this context, this book addresses the challenges associated with the explainable artificial intelligence for cyber security by providing a bigger picture of the core concepts, intelligent techniques, practices, and open research directions in this area. Additionally, the book will serve as a single source of reference for acquiring knowledge on the technology, process, and people involved in the next-generation artificial intelligence and cyber security.

Chapters

Chapter 1: The Past, Present, and Prospective Future of XAI: A Comprehensive Review
Chapter 2: An Overview of Explainable Artificial Intelligence for Cyber Security
Chapter 3: Artificial Intelligence: Practical and Ethical Challenges
Chapter 4: Domain Knowledge-Aided Explainable Artificial Intelligence
Chapter 5: Machine Learning Based IDS for Cyberattack Classification
Chapter 6: Artificial Intelligence for Cyber Security: Performance Analysis of Network Intrusion Detection
Chapter 7: Leveraging Artificial Intelligence Capabilities for Real-Time Monitoring of Cybersecurity Threats
Chapter 8: Network Forensics in the Era of Artificial Intelligence
Chapter 9: Obfuscation-Based Mechanisms in Location-Based Privacy Protection
Chapter 10: Intelligent Radio Frequency Fingerprinting to Identify Malicious Tags in the Internet of Things
Chapter 11: Explainable Artificial Intelligence for Smart City Application: A Secure and Trusted Platform
Chapter 12: Explainable Artificial Intelligence in Sustainable Smart Healthcare

The book reflects the intersection of artificial intelligence and cyber security. Unlike other books on similar topics, the book focuses on the 'explainability' of cyber security applications. Chapter 1 showcases a holistic view of explainable artificial intelligence, Chapter 2 dives into cyber security using artificial intelligence. Chapter 3 highlights ethical issues associated with artificial intelligence. Chapter 4 focuses on domain-knowledge aided explainability. Chapters 5–7 focus on network intrusion detection in depth. Chapter 8 includes insights on network forensics. Chapter 9 discusses privacy preservation and Chap. 10 highlights malicious tags identification for the Internet of Things (IoT). Chapters 11 and 12 showcase different applications of explainable artificial intelligence on smart cities and healthcare systems.

Joondalup, Australia   Mohiuddin Ahmed
West Hartford, USA   Sheikh Rabiul Islam
Melbourne, Australia   Adnan Anwar
Campbell, Australia   Nour Moustafa
Dhaka, Bangladesh   Al-Sakib Khan Pathan

Acknowledgments

It is another incredible book editing experience and our sincere gratitude to the publisher for facilitating the process. This book editing journey enhanced our patience, communication, and tenacity. We are thankful to all the contributors, reviewers, and publishing team. Last but not least, our thanks to our family members whose support and encouragement contributed significantly to completing this book.

Mohiuddin Ahmed
Sheikh Rabiul Islam
Adnan Anwar
Nour Moustafa
Al-Sakib Khan Pathan

Contents

Chapter 1
The Past, Present, and Prospective Future of XAI: A Comprehensive Review

Muhammad Usama Islam, Md. Mozaharul Mottalib, Mehedi Hassan, Zubair Ibne Alam, S. M. Zobaed, and Md. Fazle Rabby

Abstract With the increasing growth and availability of data, Artificial Intelligence (AI) based black-box models have shown significant effectiveness to solve real-world and mission-critical problems in a wide range of domains such as finance, healthcare, security, natural language processing, transportation, military, and retail. However, the importance of interpretable and explainable decision-making processes within AI-based systems is becoming crucial to provide transparency and confidence among end-users from diverse backgrounds. Acceptance of these black-box models considerably depends on to which extent the users or technical personnel understand and trust the underlying mechanism. However, ML methods and models are getting more sophisticated and transparent progressively. Though the domain experts understand the mathematical principle, they still face difficulties expressing the mechanism for a wide audience. To address this issue, a recent research domain—Explainable AI (XAI), has emerged intending to generate the contextual explanatory model for practical deployment. In this chapter, we systematically review and examine the existing literature and contribution of XAI in different domains. The presented taxonomy contains our investigation on the necessity of explaining and contextualizing

M. U. Islam (✉) · M. Hassan · S. M. Zobaed · Md. Fazle Rabby
University of Louisiana at Lafayette, Louisiana, LA 70503, USA
e-mail: usamaislam1@louisiana.edu

M. Hassan
e-mail: mehedi.hassan1@louisiana.edu

S. M. Zobaed
e-mail: sm.zoabed1@louisiana.edu

Md. Fazle Rabby
e-mail: mdfazle.rabby1@louisiana.edu

Md. Mozaharul Mottalib
University of Delaware, Newark, DE 19716, USA
e-mail: mmmdip@udel.edu

Z. I. Alam
York University, Toronto, ON, Canada
e-mail: zalam48@yourku.ca

© The Author(s), under exclusive license to Springer Nature Switzerland AG 2022
M. Ahmed et al. (eds.), *Explainable Artificial Intelligence for Cyber Security*,
Studies in Computational Intelligence 1025,
https://doi.org/10.1007/978-3-030-96630-0_1

methods and models in the AI ecosystem. Moreover, our study includes prospects in this field and the possible consequences of more explainable and interpretable ML models concerning different domains. This critical taxonomy provides the opportunities and challenges in the area of XAI that serves as a reference to future AI researchers.

Keywords Explainable AI (XAI) · Interpretable AI · Applications · Black-box models

1 Introduction

The term "AI" coined in 1956 can be defined as a system's ability to correctly interpret, to learn from external data, and to leverage the learning to achieve specific objectives through flexible adaptation [1]. In today's world, AI-based solutions are used to solve large-scale complex computational tasks with unprecedented performance efficiency. AI is significantly applied in numerous use cases with enormous consequences for humans [2–8]. From intelligent agents such as Google Home or Siri, intuitive next word prediction offered by smartphone's keyboard (e.g., GBoard, iOS keyboard) to the sophisticated computer vision models used for autonomous driving, AI is implemented. Such diversified use cases vividly highlight the widespread of AI. Hence, the credit towards the recent rapid growth of technological advancement goes to artificial intelligence. Basically, the underlying model complexity is increasingly growing which makes the AI applications smarter. On the contrary, because of the growth of the model and algorithmic complexity, the internal learning process as well as the model prediction analysis is not entirely explainable and raises questions regarding the workflow of AI capabilities.

The vague understanding of AI leads an individual to be perturbed by the actions and outputs of AI models and so, they might be considered as a sort of "black box" to most of us. The trade-off between performance and explainability can create an impact on individuals, businesses, and society. AI interpretability and explainability (XAI) have become a popular research topic in recent days to AI research community [9–11]. The neural network models are often considered as black-box models where increasing the model complexity offers unfathomable capabilities in the form of high prediction accuracy while performing various computational tasks. Because of such advancements, black-box models are significantly employed to make crucial predictions in various critical contexts [12]. The problem escalates on making and leveraging decisions that are not justifiable, rational, or that is not explainable [13].

Recently, the advancement of sophisticated AI-based systems make human intervention obsolete. There is an emerging need to understand and interpret the decisions or predictions made by the AI-based systems [14]. For instance, explanations or interpretations regarding the output of an AI system are imperative in the domain of precision medicine. The real-time decisions in autonomous vehicles, legitimate traffic behavior as part of the network security, transparency in decisions of a financial company can have an enormous effect on human lives or on their activities. The introduction and ever-increasing usage of Explainable AI (XAI) is solving the problem of explainability, understandability, and interpretability of sophisticated black box AI systems. XAI is revolutionizing different ML techniques by producing more explainable models while maintaining predictive accuracy. Transparency in system design has enabled the end-users to understand and trust the decisions of the AI systems. It also helps system designers to figure out the weakness of their model and improve predictive accuracy. Moreover, the post-hoc explanations provide justification of why any particular decision has been given to the users, more insight about every decision, or visualizations to better interpret the AI systems.

In this chapter, we discuss the current trend of Explainable-AI (XAI) in different domains. More specifically, we provide a comprehensive discussion of the performance and explainability of XAI compared to AI in Sect. 2. Then we present how XAI has evolved over the years in Sect. 3. Next, we categorize applications based on their feature importance from various domains in Sect. 4. Additionally, we discuss how XAI is increasing trust among the users of its AI-based decisions in Sect. 5. We identify the challenges and opportunities of XAI in Sect. 6. Finally, we conclude the chapter in Sect. 7.

2 XAI Versus AI

The concept of AI can be traced back to the 1940s when McCullough and Pitts proposed the development of the first neural network [15]. It was in the late 50s that the term AI was mentioned, the scope of AI was defined and a new paradigm was set [16]. In that time, AI gained substantial momentum due to the introduction of the Turing test [17] which set forth the tone of AI that we observe today. Although various formal definitions of artificial intelligence (AI) has been attributed and contributed by several researchers and entities [18], we define AI in the simplest way possible-

AI is a sub-branch of computer science that deals with building smart and intelligent systems that are comparable to human intelligence.

If we delve into the definition, we observe that AI itself is not an entity but a sub-entity of greater research related to computer science which sets the tone for where AI research should be included and headed. The second pointer is to discuss what is meant by dealing with building smart and intelligent systems. The goal and scope of AI are set to build intelligent systems. The intelligent systems are built under the scope of reactive machines, limited memory, theory of mind, and having self-awareness. In the next segment, the term "comparable to human intelligence" is further stretched to describe that there are various tasks that usually require human intelligence such as visual perception, understanding speech, decision making, policy making, word translation to quote a few which nowadays can be done by AI through to image processing, image recognition, speech recognition and synthesis, recommender systems, neural machine translation and so forth.

The next point of AI is the process that AI utilizes to make such human-like decisions. AI relies on thinking humanly and rationally and acting humanly and rationally on the basis of scope pointed out above.

In the scope of reactive machines, they have the capability of using intelligence on the basis of what is in present or in front of them with visual sustenance but doesn't react based on past as it doesn't store any form of memory. Famous examples of reactive machines included IBM's Deep Blue that defeated Gary Kasparov in the chess championship [19] and Google's AlphaGo which claimed a 4-1 victory against Go champion Lee Sedol [20]. To summarize,reactive machines are good at attaining a certain level of complexity, reliability, and can operate on tasks that can be replicated and repeated.

Limited memory, in contrast to reactive machines, can store past memory to generate intuition, insights and take sustainable decisions for the future which makes them more promising and reliable than that of reactive machines. Some popular example includes reinforcement learning, long short term memory, and generative adversarial networks. The other concepts, theory of mind and self-awareness are still at theoretical conceptualization and infancy stages where a network is fed enough data to produce innovative and new theory with self-aware capabilities.

AI as an umbrella term is currently used with various variations of subsets ranging from machine learning, deep learning, neural networks, data science which are typically used for speech recognition to image processing to search engines to self-driving cars. Popular examples today include Siri, Alexa, Amazon Rekognition, Google search, DeepMind, Uber self-driving cars, etc.

Explainable Artificial intelligence (XAI) on the other hand is a subset of AI that deals with the explainability of AI which emerged into its own field of research in the twenty-first century due to increasing concerns from stakeholders of AI. The stakeholders of AI consist of domain experts, researchers, corporate policymakers, field-level executives and managers, regulatory agencies, and the users affected by the decision. A complete visualization of stakeholders of XAI is provided in Fig 1.

The aim of XAI is to create explanations of certain decisions that are understandable by laypersons and trustworthy while maintaining a higher level of performance. A comprehensive step-by-step comparison between AI and XAI can be visualized in Fig. 2.

In brief, The process of explanation and human comprehension in the loop of AI is defined as XAI. The reason XAI became a new field of research is because of its growing interest from stakeholders in terms of research, applicability, accessibility, usability, and user acceptance. The following section would discuss in detail the evolution of XAI and the state of research where XAI is currently.

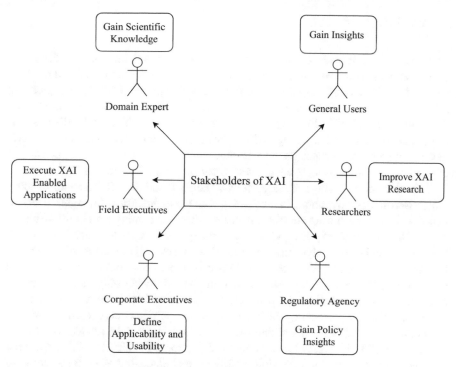

Fig. 1 Figure outlining the important stakeholders of XAI along with associated expected outcome from stakeholders

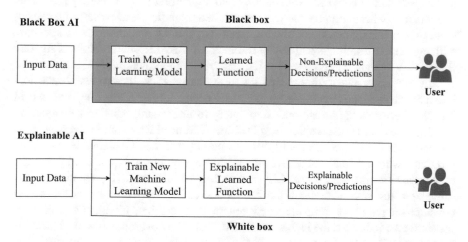

Fig. 2 A figure for visualizing a clearer contrast between XAI and AI

3   Evolution of XAI

The much-hyped term in present era Explainable-Artificial Intelligence (XAI) can be traced back to the root and core researches produced by Scott [21] where they deduced the importance of explaining certain decisions via a rule-based approach taking into consideration the variables that are in play. Multiple research work by Scott and Shortlife [22–24] confirms that the building blocks of XAI albeit a new term in the present era has a long history. We are intrigued to find such a hyped term has a root that almost dates back to 45 years of research where the importance of explaining insights was realized at a very early stage of AI research [22].

The AI in its early stage of research was mainly based on rule-based approaches which were defined by human beings which made explainability of insights an easier task than that of the current era where the researchers are dealing with multiple black-box model structure [25]. A good example of explainability is the decision tree that has an inherent explainable infrastructure that makes explainability of a decision tree an easier task [26]. The data explosion of the twenty-first-century followed by the popularity of complex machine learning and deep learning models started a new revolution of paradigm for researchers in terms of better metrics in accuracy, precision, recall outperforming the older models of approach thus eventually making the complex neural networks an industry-standard in any tasks related to artificial intelligence [27]. What the research community ignored to address in the last decade was the adverse reduction in explain-ability of such decisions with the increase of complex black box models [28]. The adverse reduction can further be characterized as an inverse relation between explainability and higher prediction accuracy [29]. This phenomenon where complex model produces better prediction accuracy with respect to near-zero transparency and simple model performs worst in terms of accuracy with greater explanation is pointed by Defence Advanced Research Projects Agency (DARPA)'s explainable AI program [26, 29, 30] and redrawn in Fig. 3.

The tone of explanation that was set forth almost 45 years ago which was lost into data-chaos in the twenty-first century again came into attention due to perceived breach of privacy, user trust, and users query to understand insights generated by AI [31]. The continued attention lead Defense Advanced Research Projects Agency (DARPA) in the United States in 2017 to commence the "Explainable AI (XAI) program" whose primary objective was set to improving the explanation paradigm of AI decision making process [29, 30] with a 50 million USD incentive. In the same year, the sino-us competition for AI supremacy led China to release a new development plan for a completely new generation of artificial intelligence where the research plan indicated the inclusion of XAI [26]. The government-backed sensible turn towards XAI led Europe to adopt the "general data protection regulation" The following year [32] where the right to an explanation was included in the case of personnel who were influenced and affected by AI induced decisions. The decision to shift towards transparency of AI by three major players in the world eventually led XAI to be accepted globally as a key factor of importance and emergence [33]. This specific need for transparency in AI lead all the stakeholders to seriously make

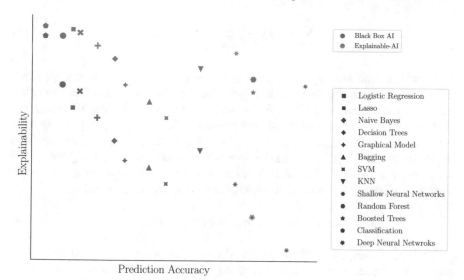

Fig. 3   Figure depicting inverse relation between accuracy and explainability of AI models

a pragmatic turn to explore the need for XAI in the current world [34] where a
positive tone for the explanation of decisions, insights, and recommendations was
emphasized [9] thus making the inclusion of XAI in AI a new-normal.

As research started to progress in XAI, In an attempt to open the black box of
currently existing deep neural networks, two conclusions were reached in terms
of achieving explainability namely transparency design and post-hoc analysis [26].
The transparency design deals with how exactly a model operates as it goes along
training whereas post-hoc analysis explains why a certain result has been concluded
as a recommendation to the user.

Currently, various ways are being explored to induce explain-ability in AI [35].
One such example is a very old concept called Shapley values that estimates the
contribution of a feature value to the prediction in different coalitions [36]. In mathematical terms, Let there is a linear model, $f(x)$ such that,

$$f(x) = W_0 + W_1 x_1 + W_2 x_2 + \ldots \ldots W_n x_n \tag{1}$$

where, contribution C_i of the ith feature is on the prediction $f(x)$ is, so

$$C_i(f) = W_i x_i - E(W_i \times X_i) \tag{2}$$

where, $E(W_i \times X_i)$ is the mean effect estimate for feature i which means this mathematically determines the feature contribution that if we sum for all features for one
instance would lead to the following equation.

$$\sum_{i=1}^{n} C_i(f) = f(x) - E(f(X)) \tag{3}$$

The problem with Shapley value estimation remains in the time-hungry approach to estimation. Similar methods that explore feature importance and feature contribution can be found in [37–39]. Furthermore, the exponential size of estimation with feature expansion is addressed through Monte-Carlo sampling in [38]. Another robust alternative to Shapley value estimation is Local interpret-able model-agnostic explanations(LIME) [40]. The mathematical explainable model (XAI_l) for LIME for instance x, model m that minimizes loss S that measures explanation f for the model complexity $V(g)$ with G possible explanations can be deduced as:

$$XAI_l(x) = argmin_{g \subseteq G} S(f, m, \pi_x) + V(g) \tag{4}$$

However, LIME, unlike Shapley value, does not guarantee fair distribution in their model. An alternative can be found in SHAP [41] which implements Shapley value but the concept of "always use all features" as idealized in shapley is not taken into consideration in SHAP. The discussion yields a simple connotation. There would always be a trade-off with time versus feature use versus explainability as using more features to explain would take more time to generate explainable values. One example of transparency design can be found in research by Zhang and his team [57] that uses feature maps to generate explanations. Another widely used process is called layer-wise relevance propagation (LRP) [52] where through a full decomposition it analyses the pixel contribution to prediction as opposed to sensitivity analysis [53] that encompasses partial derivatives instead of full decomposition.

Mathematically, Sensitivity analysis that measures relevance score on locally evaluated gradient can be defined for RS_i as;

$$RS_i = \left\| \left\| \frac{d}{dx_i} f(x) \right\| \right\| \tag{5}$$

The LRP in contrast to SA explains the full decomposition to provide information of all the pixels available and their relevance in the decision by generating a heatmap.

This discussion brings us towards another door of questions that need to be asked.

– Which models are interpretable already (to some and/or full extent)?
– What are some approaches of model agnostic methods in the current state?
– How to interpret neural networks?

Linear regression, logistic regression, generalized linear models, generalized additive models, decision trees, naive bayes classifier, kNN models, are some of the wide variety of examples whose decisions can be interpreted mathematically upon which can be described in lay-person terms [58].

We have identified several approaches of model agnostic methods as well as interpretation approaches [59] for neural networks as well as black-box models currently in use which we summarize in Table 1.

Table 1 Outlining several currently used approaches to XAI along with remarks pertaining to each approach

References	Approach	Remarks
[42]	Partial Dependence Plot (PDP)	Shows feature effect on outcome
[43]	Individual Conditional Expectation (ICE)	Shows instance prediction change with feature change
[44]	Accumulated Local Effects (ALE)	Plot showing how feature influence prediction
[45]	Feature interaction	Measures interaction strength of features
[46]	Functional decomposition with functional ANOVA	Decomposing complex ML into individual effects and interactions
[47]	Permutation Feature Importance	Shows increase in prediction error after feature value permutation
[40]	Local interpretable model-agnostic explanations (LIME)	To approximate the outcomes of black box model
[48]	Anchors	Rule-based decision (Anchors) to explain predictions
[41]	SHapley Additive exPlanations (SHAP)	Calculates feature contribution
[49]	Feature visualization	Making learned features explicit
[50]	Network dissection	Quantifies the interpretability of a unit of CNN
[51]	Pixel contribution	Highlights relevant pixels in image classification tasks
[52]	Layer-wise relevance propagation (LRP)	Backward Pixel decomposition to generate relevance
[53]	Sensitivity analysis	Partial derivative and relevance score
[54]	Concept activation vectors	Measures influence of a concept
[55]	Concept whitening	Interpretable concept generation
[56]	Concept bottleneck models (CBM)	Mapping inputs to concepts

It is evident that currently, XAI uses feature importance, heat map, activation values, feature contribution, quantifying feature importance through the weighted scoring matrix to generate explanations of which some early and latest works were put into contrast in the aforementioned discussion to understand the evolution of XAI.

4 Current Applications of XAI

Recent years have seen an XAI-focused research proliferation that eventually elicits massive applications and deployments. Feiyu et al. discuss the important group of stakeholders in this massive endeavor [26]. The major three groups are the users who utilize the AI systems, the people who are affected by the AI decisions, and the developers of the AI systems and algorithms. For example, a medical practitioner needs to understand the features in the input data which lead to the AI-generated diagnosis reports. In this case, the medical practitioner is a user of the system and the understanding of the high-level intricacies at a minimal level is very detrimental for the system. Then, there comes the people who are affected by AI decisions. Already established legal regulations are there that codify the demand of the affected peoples' need to understand the reason behind any inference. If the AI model makes its own decision, e.g., diagnosing a condition, braking the car, or issuing a punishment order for a crime, the affected people have all the right reasons to get the explanations. Finally, the group of developers of AI models who could help to improve AI systems and algorithms has a significant share in this issue of explanations. Lapuschkin et al. [60] presented an interesting example. They figured out despite showing lower accuracy by the Fisher Vector method in the task of object recognition compared to the Deep Neural Networks, the Fisher Vector method does a brilliant job at recognizing the category "horse" which is very much comparable to the other method. On analyzing the models using a saliency map method named "Layer-wise Relevance Propagation" [61] for visualizing the pixel areas emphasized by the models, they observed that the Deep Neural Network model looked at the contour of the actual horse whereas the Fisher Vector model mostly relied on a certain copyright tag that happens to be present on many horse images. This example elicits how XAI plays a very important role in improving state-of-the-art systems and algorithms. To discern the applications of XAI, we will have to look at different sectors where XAI applications are gaining significant advancements and distinguish the groups concerned with the models. In all the cases, the group that helps in improving the systems or algorithms seems to be the developers of the model. On the other hand, groups being affected and groups who utilize the system are different in different scenarios and sectors.

4.1 Healthcare

Healthcare is one of the best examples in which the achievements of AI have shown the greatest practical value. Specifically, AI has revolutionized the field of a cancer diagnosis. Besides that, discovering new uses of existing drugs, revealing cancer in tissues, detecting cardiac arrhythmia, and predicting hypoglycemic events in diabetics, etc. are some of the examples that show the attachment of AI to healthcare system enabled efficiency and advancement. AI applications ranges from imputing missing data in Electronic Health Records (EHR) [62–64] to clinical outcome prediction (e.g., Diabetes prediction [65–67] and patient monitoring, image-based diagnosis [68, 69], genome interpretation and biomarker discovery [70, 71], inferring health status through wearable devices [72], autonomous robotic surgery [73].

As AI has become an active part of health care systems, the increasingly widespread applications of these models entail the need for explanations to hold such models accountable. Moreover, in health care, critical applications such as the prediction of a patient's end of life or developing cancer may have much more uncompromising conditions for explanation fidelity as compared to predicting merely the costs of a medical procedure. The demands for explainability, model fidelity, and performance are generally much higher in healthcare as compared to most other domains which give rise to unique challenges. For ease of discussion, we will be dividing XAI applications in healthcare into two parts, the first one using traditional ML techniques and the latter one having more robust techniques such as Deep Learning.

Using the self-explanatory surrogate models of Decision Tree (DT) simple if-then rule explanations of the decisions can be acquired. Al-Dlaeen et al. implemented DTs to predict Alzheimer's disease using gender, age, genetic causes, brain injury, and vascular disease as data attributes [74]. They measured the information gain of attributes for the selection of nodes. A hybrid approach of incorporating both the benefit of individual DT and competence of ensemble DTs, Gibbons et al. developed a Computerized Adaptive Diagnostic (CAD) system for the diagnosis of major depressive disorder (MDD) using random forest and DTs [75]. Suresh et al. proposed the use of radial basis function network and DTs for detecting lesions in mammograms [76]. Khaere et al. uses the association rule technique to detect heart diseases using 23 attributes of cardiovascular data [77]. For automatic answering to healthcare-related questions, Agrawal et al. [78] implemented a rule-based system for the classification of questions. The rules were extracted pre-processing 427 health-based questions to identify 9 types of questions. For better accuracy and performance, an ensemble of DTs is preferred over a single DT, therefore random forest (RF) is one of the most common algorithms used for regression or classification with explanations. Wang et al. implemented an RF model for the detection of epileptic seizures and classifying based on Electroencephalogram (EEG) signals [79]. For the simplicity and self-explanatory structure of RFs, these models are used for predicting the depression of Alzheimer's patients [80], evaluating Covid-19 severity [81], healthcare monitoring systems [82], prediction of medical expenditures [83]. Another simple ML regression model, General Additive Model (GAM) is being extensively used for health-related environmental research [84, 85], pneumonia risk prediction [86], distribution of species and the effect of age, and a diagnosis-specific cohort of HIV patients on psycho-social and behavioral activities [87] (Tables 2 and 3).

Several applications involve model-agnostic explanations such as partial dependence plot (PDP), class activation map (CAM), LRP, etc. Yang et al. [88] predicted the mortality of COVID-19 patients using age, time in hospital, gender, and chronic disease as attributes. Using PDP they concluded that the age of a patient is the most important factor along with the time in hospital being the next significant indicator of mortality. Gupta et al. [89] used VGG16 to detect acute proximal femoral fractures in the elderly, CAM helped them to localize the fractures. Kumar [93] et al. proposed a mosquito net to detect malaria cells and explained the decisions using the Grad-CAM (a variant of CAM). Goodfellow et al. [90] used CAM to evaluate errors of their CNN model proposed for the multiclass labeling of ECG signals. Pereira et al. [91] proposed a CNN classifier for brain tumor detection and used CAM to explain the brain tumor grading decision. Irvin et al. [92] used the GradCAM to deliver the

Table 2 Summary of model-intrinsic explanation methods for healthcare applications

Explaining method	Description	References	Application	Modality
Decision Trees	The self-explanatory surrogate model uses if-then logic for decision	[75]	MDD detection	Psychiatric and non-psychiatric attributes
		[74]	Alzheimer's disease detection	Gender, Age, Genetic causes, Brain injury, vascular disease
		[76]	Breast cancer detection	Mammographic images
Rule-Lists	Textual format explanation using if-then logic for decision making	[77]	Cardiovascular disease detection	Various attributes of patients
		[78]	Question classification in health care	Coarse and fine-grained classes from cloud questionnaire
Random Forest	An ensemble of large numbers of DTs, used mainly for regression or classification	[79]	Epilepsy detection	EEG signals
		[80]	Alzheimer's patients' depression detection	Social demographic factors, Health status, behaviors, living style, economic activity
		[82]	Healthcare monitoring system	Breast cancer, Diabetes, Heart disease, spect-heart, thyroid, surgery, dermatology, liver disorder
		[81]	Covid death and recovery rate detection	Categorical variables in dataset such as fatigue, fever, cough.
GAM	The output is modeled as the weighted sum of random nonlinear functions of data features	[86]	Pneumonia risk prediction	Various attributes of patients
		[87]	Effect of age and diagnosis-specific cohort of HIV patients on psychosocial activities and behavioral activities	Various attributes related to psychosocial and behavioral outcomes
		[84]	Effect of air pollution on pregnancy	Various air pollutants data
		[85]	Study air pollution effect on TB cases	Pulmonary TB and air pollutants data

Table 3 Summary of model-agnostic explanation methods for healthcare applications

Explaining method	Description	References	Black box model	Application	Modality
PDP	Highlight the partial effects the input features have on the prediction of a black-box model	[88]	XGBoost	Mortality rate in COVID-19	Age, gender, time to hospital
CAM	Highlight the class relevant areas of input data.	[89]	VGG16	Fracture detection	X-Rays
		[90]	CNN	ECG classification	ECG signals
		[91]	CNN	Grading of brain tumor	MRI
GradCAM	Generates weighted gradient CAM by computing gradients of output as it goes towards last layer	[92]	CNN	Detection of different diseases	Chest X-Rays
		[93]	CNN	Malaria detection	Blood samples
MLCAM	Generates the maps of discriminating features of data	[94]	CNN	Brain tumor detection	MRI
LRP	Back-propagates the output decision to the input layer to estimate the relevance of each attribute.	[95]	LSTM	Cancer therapy decision prediction	Various
		[96]	Semantic segmentation network	Liver tumor classification	MRI
LIME	Decompose the data based on similar features and tweak randomly selected features to measure output dependence	[97]	CNNVGG16	Detection of metastases	WSI patches
		[98]	CNN	Detection of metastases	Immuno fluorescent images
DeepLIFT	Uses a reference value and measures the reference values of all neurons using a forward and backward pass	[99]	–	Genetic variants caused by diseases	Single-nucleotide polymorphisms
SHAP	Uses coalitional game theory to calculate Shapley values that show the distribution of prediction among features	[100]	LR, SVM, RF, XGboost, RF+XGboost	Detection of cardiac surgery-associated acute kidney injury	Various disease related features
GBP	Backpropagates the positive error signals by setting negative gradients to zero and limits itself to positive inputs	[101]	3D-CNN	Detection of Parkinson's disease	SPECT
AM	Generate explanations by maximize the activation of neurons tweaking the input	[102]	CNN	Detection of hip implant misplacement	X-rays

visual explanations of active pleural effusion areas of chest radiograph indicated by a CNN model. Izadyyazdanabadi et al. [94] integrated multiscale activation maps (MLCAM) with the CNN model to locate the features of glioma tumors. Yang et al. [95] used LRP to select the features with high relevance for predicting the decision of therapy of patients. Chlebus et al. [96] implemented an LRP algorithm to explain the decisions of semantic networks used for liver tumor segmentation. De Sousa et al. [97] implemented LIME for generating the explanation of how CNN and VGG16 detect the metastases from the histology WSI patches. Zafar et al. [103] proposed deterministic LIME (DLIME) for explaining the decision of three medical domains i.e., breast cancer, liver disease, and hepatitis detection using hierarchical clustering and K-nearest neighbor classifier. Kitamura et al. [98] detected diabetic nephropathy using a CNN network from the immunofluorescent images and analyzed the decisions of CNN using LIME. Sharma et al. [99] implemented a Deep Learning Important FeaTures (DeepLIFT) for studying genetic variants caused by common diseases. DeepLIFT was able to reveal diabetes genetic risk factors using DL techniques. Tseng et al. [100] used SHAP values to describe the feature importance of multiple features in studying the cardiac surgery-associated acute kidney injury. These factors are not revealed using traditional techniques. Yu et al. [104] showed the contribution of each plasma feature in developing lung cancer using SHAP. Couteaux et al. [105] implemented sensitivity analysis of each feature by maximizing the neuron activation using gradient ascent in a study of the classification of tumors using data of liver computed tomography. Pianpanit et al. [101] showed that guided backpropagation performs best while explaining the 3D-CNN architecture for Parkinson's disease detection. Borjali et al. [102] implemented activation maximization (AM) to visualize the classification of hip implant misplacement using X-rays dataset for orthopedic application.

4.2 Industry

AI is playing an integral role in determining our everyday experiences. The abundance of AI-based solutions in areas such as job recruitment, loan sanctioning, criminal justice, or education has made the impact of AI extensive in personal and professional sectors. The dominant role of AI in these domains has led to a growing concern regarding potential bias in these models and demand for model transparency and interpretability [30]. Moreover, model explainability is a precondition for building trust and adaptability of AI systems in high-stakes domains requiring reliability and safety. In industry, several applications of XAI can be found which ensures these models are scalable as well as interpretable (Table 4).

Wrangler and OpenRefine offer customizable data transformations facilitating the data mining process. Wrangler offers an interactive system for creating data transformations combining direct manipulation of visualized data with the automatic inference of relevant transforms. Analysts are enabled to iteratively explore the space of applicable operations and preview their effects [106]. In OpenRefine

Table 4 Summary of XAI based applications in industry

References	Approach/system	Remarks
[106]	Wrangler	Reorganize data in order to create easy to understand views
[107]	OpenRefine	Explore, clean, transform and reconcile data with various web services
[108]	Tensorflow embedding projector	Interactive visualization and analysis of high-dimensional data using ML techniques
[109]	Debugging tool	Help AI engineers better understand ML models and debug issues
[110]	Ensemble Decision Trees	Predict credit risk with rules
[111, 112]	Markov Model	Predict discrimination and rule inference
[113]	Gradient boosting, DL	Credit risk analysis
[114]	AutoAI	Project management curriculum

users can explore data to see the big picture, clean and transform data, and reconcile data with various web services [107]. While OpenRefine cleans data, Wrangler provides means to reorganize data in order to create easy-to-understand views. However, both the systems lack at providing any data quality parameters or their explanations. They depend on the user to drive their own ML processes to generate and interpret the views [115]. Tensorflow Embedding Projector moves a step forward as it offers an advanced tool for interactive visualization and analysis of high-dimensional data using ML techniques such as uniform manifold approximation and projection (UMAP), t-distributed stochastic neighbor embedding (TSNE), and principal component analysis (PCA). It provides multiple dimensions of interpretation: Exploring local neighborhoods, viewing global geometry, and finding clusters [108]. Qui et al. [109] provides two debugging tools that provide for search and feed to visualize the relevant information. This helps AI engineers better understand their machine learning models and debug issues.

Florez-Lopez et al. [110] adopt an ensemble of decision trees and explain predictions of Credit risk with rules. Volkov et al. [111] implemented Markov models for discrimination. Markov models are also used in rule inference by Xu et al. [112]. Addo et al. [113] use a number of black-box models, including gradient boosting and deep neural architectures. XAI is also being incorporated in project management (PM). The current teaching methods are developed decades ago and require AI tools to automate them. Though the advent of AI and its use in PM provides new opportunities for prediction and better results, systems with explanations are scarce. Radhakrishnan et al. [114] introduced AI methods with explanations to be used in IBM's Watson AI platform.

4.3   Security

With the advancement of machine learning-based network intrusion detection systems, the Cybersecurity domain has started to see effective results in recent times. To bypass these guards, the intruders are now attacking with new patterns. With the help of the applications of XAI, researchers have come up with preventive solutions. For example, the data-driven Intrusion Detection Systems (IDSs) often fail when the model is exposed to a new attack where the data is unavailable and hence that is misclassified as normal resulting in a system breach. Therefore, having a clear explanation of such causes of miss-classification is necessary to prevent further attacks. The minimum modifications required for correctly classifying the miss-classified samples can be determined using the adversarial machine learning approach [116]. This is achieved by utilizing the generated explanations for erroneous classifications caused by the data-driven IDSs and can be further extended to execute vulnerability assessment.

Explanations of the reasons why a specific prediction is made provide us with the features that play a vital role in predicting cyber-attacks. Data scientists understand what patterns the model has learned and can modify the dataset accordingly if the learned patterns are problematic. Deep neural network-based intrusion detection system with embedded XAI framework [117] provides transparency at each level of learning and helps the data scientists to get a much clearer picture. The explainability increases with the infusion of popular network security principle (CIA Principle) [118] into the AI model. Such systems show faster response against unknown malicious attacks as well (Table 5).

Auto-detection of alarm labels populated in security information and event management (SIEM) systems and associating them with different attack types have been made possible by applying the zero-shot [119] learning method to machine learning by leveraging the interpretability of predictions generated by a model. Using XAI,

Table 5 Summary of XAI based security applications

References	Approach	Remarks
[116]	Adversarial machine learning	Utilizes the generated explanations for wrong classifications
[117]	Deep neural network	Provides transparency at each level of prediction
[118]	Infusion of CIA Principle	Shows faster response against attacks
[119]	Zero-shot learning method	Attacks can be addressed without prior knowledge
[120]	Clustering	Supports unsupervised data analysis
[121]	Shapley–Lorenz decomposition	Evaluate performance based on rank regression and Shapley-Lorenz decomposition
[122]	Decision tree	Identify malicious traffic

an attack without prior knowledge can be identified. Another example of XAI-based methodology to understand the reason that leads to a certain prediction made by a supervised model is shown in EXPLAIN-IT [120]. The introduction of model agnostic and powerful Shapley values was made to understand why a prediction has been concluded by a model. Unfortunately, it could not become a standard procedure since the Shapley values are not normalized. XAI model [121] embedded with statistically normalized Shapley values helps assessing and identifying the key drivers of cyber risk. The XAI concept has been addressed [122] to enhance cyber risk management by exploring a popular decision tree model that replicates the human approach of decision making. By interpreting the decision tree rules exposed by XAI, network security personnel can take an effective course of action to identify malicious traffic.

4.4   Natural Language Processing

The Natural Language Process (NLP) domain is enriched with various complex, black box Deep Learning models. In recent years, NLP research has also focused on the explainability or interpretability of AI-based models. The explanations can be categorized based on multiple aspects such as Local vs Global, Self-Explaining versus Post-Hoc explanations. Additionally, recent studies use several explainability techniques including feature importance, surrogate model, example-driven, induction, and provenance.

Wallace et al. used leave-one-out conformity as a metric for model uncertainty coupled with saliency map as a visualization technique for text classification [123]. Poerner et al. evaluated different explanation methods on Small and Large Context NLP tasks [124]. To explain the reasoning capabilities, Jiang et al. proposed multiple reasoning-chain recovery tests [125]. Ghaeini et al. provided an interpretation technique of the natural language inference (NLI) exploiting the saliency of attention and gated signals of LSTM [126]. Dong et al designed interpretable the neural programmer-interpreter (NPI)-based model by providing operation traces [127]. QUINT is a live question answering system [128]. It provides a complete derivations sequence along with the visualization for explainability and interpretability. Pryzant et al. relied on class activation mapping, attention, and gradient reversal for explainability of the performance of ads on different writing styles. Pryzant et al. [129]. Liu et al. utilized adversarial perturbation for interpretation in-network embedding [130]. Pröllochs et al. proposed inferred negation rules to provide interpretability in negation scope detection [131]. Studies in [123–125] provide local explanation or justification of the exploited model's prediction following a post-hoc explaining approach. On the other hand, [126–128] follow the self-explaining local explanation approach. While [131] provides global explanation with self-explaining approach, [129, 131] follow global explanation with post-hoc approach. XAI applications in NLP along with their explainability technique are summarized in Table 6.

Table 6 Summary of XAI-based applications in NLP

Category	References	Approach	Remarks
Local and post-hoc	[123]	Feature importance	Used Leave-one-out conformity as a metric for model uncertainty
	[124]	Surrogate model	Evaluated different explainable methods
	[125]	Example driven	Used multiple reasoning-chain recovery tests
Local and self explanatory	[126]	Feature importance	Used attention and LSTM gating signals
	[127]	Induction	Utilized interpretable NPI-based model
	[128]	Provenance	Provides complete sequence of derivations of questions-answering
Global and post-hoc	[129]	Feature importance	Used class activation mapping, attention, and gradient reversal on classifying different writing styles
	[130]	Surrogate model	Utilized adversarial perturbation for network embedding
Global and self-explanatory	[131]	Induction	Used inferred negation rules

5   Usability, User Acceptance, User Trust and XAI

Several research works are carried out by researchers in usability, user experience domain related to the relation of usability, user acceptance, user trust, and explainable artificial intelligence.

Weitz and his colleagues [132] carried out a user study and noted that integrating XAI within speech recognition tasks of virtual agents has lead to an increase in user trust by the users. A similar conclusion was reached by researchers in [133] where they have mentioned significant user trust and acceptance for the case of autonomous vehicles after the inclusion of XAI. Branley-Bell et al. [134] pointed out an interesting relationship between increased understanding and increased trust which is directly proportional to each other in the case of human agents. They have also stressed the need to include a sample size of various demographic aspects to understand the full extent of justification. Authors in [135] have explored rule-based explanations and example-based explanations of AI and concluded in favor of amalgamating rule and example-based explanations to obtain the best outcome. Arun Rai in his research [136] explained the positive correlation between the explanation of prediction, user trust in understanding such prediction, and users' subsequent decision to accept or reject that prediction. A similar tone was observed in Biran and his teams [137] research work where researchers mentioned that a comprehensive insight on how the system functions will positively impact the trust of the user which is in conjunction with the researchers point mentioned in [136]. Authors in [138] have examined various theories related to attribution, power status, and anthropo-morphism, to find thathumans attribute similar pointers irrespective of where the explanations are coming from. Wang et al. [139] designed a framework for explain-ing XAI results with co-design principles. Papenmier and his team [140] researched on accuracy and its subsequent explanation fidelity and interestingly noted that the explanations having high-fidelity are very insightful and understandable to human beings albeit being faithful to the model itself whereas the explanations having low fidelity although insightful in limited capacity has impacted bias towards the model. Dazeley and his team [141] researched widely on explanations generated from XAI and concluded that current explanations are not enough and research works should be carried out on generating wider forms of explanations for greater user satisfaction and acceptance.

We can conclude by saying that several research works are carried out in terms of usability, user acceptance, and user trust enhancement related to XAI which although insightful that explainability increases user trust still needs further research in more domains and with more sample sets.

6 Challenges and Opportunities

XAI as a term albeit used widely is still at its infancy stage in terms of the develop-ment of a whole white box transparent system with a plug-n-play mode. The research, evolution, and application discussed in the earlier subsequent sections clearly por-trays that although there is hundreds of research work on ensuring transparency by improving various component of a system a full-fledged system is still afar and needs substantial contribution from various stakeholders to make XAI a reality and start a process towards shifting to Responsibly Reliable-AI (RRAI) from XAI. In this section, we discuss the challenges associated with research, challenges associ-ated with stakeholder acceptance followed by opportunities and eventually providing futuristic aspirations for XAI to grow and mature.

6.1 Challenges

The main challenge of XAI now is the development of a fully-fledged white-box system that would generate understandable lay-person explanations with substantial user acceptance. We have found out that there are standalone components of XAI currently in use and are developed by researchers although the challenge is to inte-grate all the components of XAI together to make a unified model. While this would need the aid of machine learning researchers and explainable AI researchers on one hand and on the other hand this approach needs to invite User interface, User expe-rience, usability, and human-computer interaction researchers at the same platform to collaborate because the ultimate goal of XAI is understandability of decision by human stakeholders who are mainly not apt with technology in general and the main gap remains is the bridge that needs to be connected from explainability to understandability for which these segments of researchers mentioned above should work together to bridge this substantial gap as although on one hand explainability is an important factor in research, if the decisions are not understood or usable or accepted by general mass users, the research would go in vain.

The concept of understandability on the stakeholder side leads us to explore the stakeholders of adoption, accessibility, and acceptance. One of the most important stakeholders is the domain experts such as doctors, engineers, data scientists, researchers, stock agents, insurance agents who need to understand the insights for their knowledge so that they can disseminate the insights to the lay-persons. The domain experts as an intermediary between XAI and general people for an understandable insight generation system would benefit in terms of knowledge and in terms of superior explanation.

The regulatory agency that has the power of policymaking, policy execution, legislation related to computer science in general, and artificial intelligence especially would benefit immensely from understanding the insights. The various compliance laws, policies taken to be executed that are generated through AI recommendations need to be well understood as a full-fledged white box for the policy-making body to make such recommendations. The ins and outs of AI recommendations need to be explained and understood by the members of regulatory agencies. They are an important stake at the top of the AI recommendation ecosystem because regulatory agencies are the one that sets the regulations in motion and without understanding in-depth, it would not be a feasible policy in the long run.

Similar to regulatory agencies, another important stakeholder would be top executives, CEOs, CXO's, and board members of corporate organizations. There is an amalgamation of technical and non-technical expertise that makes this stakeholder a melting pot from all aspects. These specific stakeholders are responsible for policymaking that would directly affect a vast majority of consumers as companies from fortune 500 are widely used as a business-to-business and business-to-consumer model where a large chunk of consumers and businesses are directly involved. So, adopting, accessing, and accepting XAI at this level would need a substantial push, financial burden, and positive outlook from these stakeholders to make the case a reality. Also, these specific stakeholders need to understand the concept and application of XAI first so that they can utilize them in their respective industries upon which they can include them in their policy of adoption. At the secondary level, the mid-level managers and field-level executives of these companies are important stakeholders of XAI as before deploying in-ground, they need to fully understand the ins and outs of the system. Although the top-level executives would disseminate and disperse policies related to XAI, the real work of education, maintaining, and executing XAI at the ground level would fall to this specific segment of stakeholders which is why they are a substantially relevant part of XAI understanding.

Last but not the least, the most important stakeholder of all are the direct users of XAI and who were and are affected by the decisions conveyed by the AI. The recommendation that is conveyed by AI needs to be understood, accepted, and used by these stakeholders which would make XAI usable and understandable. There should be an amalgamation of XAI and technology acceptance model to understand the usability and acceptance of the system by the general mass which is the future challenge upon making a white box XAI system. The users who are affected by the decision have three rights that we have identified: (a) right to know how the recommendation is made, (b) right to understand the recommendation (c) right to accept/ reject the recommendation. These rights would directly point to the usability and acceptability of the system. This discussion of challenges and research gaps thus leads us to the next point of realizing the opportunities that all the stakeholders can avail themselves of to make XAI a sustainable future.

6.2 Opportunities

There are ample opportunities available in research related to XAI. The first opportunity of research that we have identified is making components of XAI more robust which falls under theory development research. To define robustness, we identify that the explanations currently practiced have varying explanations and are more technical oriented and an expert is needed within the loop to describe the explanation to a lay-person. The second opportunity that we have identified, is the development of a full-fledged white box XAI model by joining the various segregated components of XAI. This research work is important because unless a fully functional XAI is developed, we are yet to understand the power XAI can disseminate. Upon development, the next research work would focus on the explainability of those models to select the state-of-the-art model. Eventually, the goal of XAI, which is to make the insights understood to general people, needs to be maintained by developing a system that would take in insights from XAI and shall describe in a lay-person non-technical manner. The amalgamation of explainability and understandability of the XAI can be achieved by enjoining UI/UX researchers, NLP researchers, and ML researchers, social science and HCI researchers as well as software engineering practitioners to develop an understandable XAI system. The aim of XAI is the restoration of trust issues which is another area of research that the researchers can focus on by enabling various studies related to user acceptance and trust. Those studies involving human agents would provide us with domain knowledge as to whether the system worked or not.

The main research direction that XAI needs to move towards is towards Responsibly Reliable-AI (RRAI) which is the future of XAI. A complete visualization of the conceptual framework of RRAI is provided in Fig. 4.

The reason we argue for RRAI is, As XAI continues to mature, there should be components such as ethics, compassion, sensibility, trust, security that needs to be involved within the XAI system to make the system more acceptable and efficient to the human agents which eventually would take a paradigm shift towards RRAI.

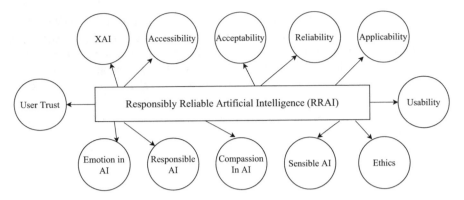

Fig. 4 A conceptual framework of Responsibly Reliable AI (RRAI) along with suggested components of RRAI

7 Conclusions

AI-based models impact human lives in many different ways. These black-box models should not be used to avoid harmful consequences that may arise from wrong decisions or predictions. Explainable approaches solve this problem by providing more insights through analysis and visualizations. In this chapter, we identified and summarized different approaches used for the explainability of the black box models. Additionally, we reviewed the current trend of XAI-based applications in healthcare, industry, security, and NLP domains. Furthermore, we discussed how explainability or interpretability, or transparency can increase trust and confidence in using XAI-based models among users. Eventually, we discussed the opportunities and challenges in different research domains using XAI. We have mentioned the critical challenges pertaining to XAI research where we stressed the need for a full-fledged white box system to be developed and proposed subsequent opportunities through findings that we have gathered. Finally, we concluded by outlining the future of XAI to be RRAI by accumulating certain humanistic components as XAI begins to mature in research and applicability.

References

1. A. Kaplan, M. Haenlein, Siri, siri, in my hand: Who's the fairest in the land? On the interpretations, illustrations, and implications of artificial intelligence. J. Bus. Horiz. **62**(1), 15–25 (2019)
2. Abdullah, F.B. Ashraf, N.S. Momo, Comparative analysis on prediction models with various data preprocessings in the prognosis of cervical cancer, in *2019 10th International Conference on Computing, Communication and Networking Technologies (ICCCNT)* (2019), pp. 1–6

3. F.B. Ashraf, M.S.R. Shafi, M.R. Kabir, Host trait prediction from human microbiome data for colorectal cancer, in *2020 23rd International Conference on Computer and Information Technology (ICCIT)* (2020), pp. 1–5
4. M. Usama Islam, M. Hossain, I. Hossain, M. Abul Kashem et al., Diabedetect: a novel decision tree-based approach for early prognosis of diabetes, in *International Conference on Innovative Computing and Communications* (Springer, 2022), pp. 563–571
5. S. Zobaed, M. Hassan, M.U. Islam, M.E. Haque, Deep learning in IOT-based healthcare applications, in *Deep Learning for Internet of Things Infrastructure* (CRC Press), pp. 183–200
6. M.T. Hoque, A. Islam, E. Ahmed, K.A. Mamun, M.N. Huda, Analyzing performance of different machine learning approaches with doc2vec for classifying sentiment of bengali natural language, in *2019 International Conference on Electrical, Computer and Communication Engineering (ECCE)* (IEEE, 2019), pp. 1–5
7. M.F. Rabby, Y. Tu, M.I. Hossen, I. Lee, A.S. Maida, X. Hei, Stacked lSTM based deep recurrent neural network with Kalman smoothing for blood glucose prediction. BMC Med Inform Decis Making **21**(1), 1–15 (2021)
8. A.I. Anik, A. Bunt, Data-centric explanations: explaining training data of machine learning systems to promote transparency, in *Proceedings of the 2021 CHI Conference on Human Factors in Computing Systems* (2021), pp. 1–13
9. A. Adadi, M. Berrada, Peeking inside the black-box: a survey on explainable artificial intelligence (XAI). IEEE Access **6**, 52138–52160 (2018)
10. S. Dhanorkar, C.T. Wolf, K. Qian, A. Xu, L. Popa, Y. Li, Tutorial on explainability for natural language processing, in *Proceedings of the 1st Conference of the Asia-Pacific Chapter of the Association for Computational Linguistics and the 10th International Joint Conference on Natural Language Processing* (2020)
11. R. Guidotti, A. Monreale, S. Ruggieri, F. Turini, F. Giannotti, D. Pedreschi, A survey of methods for explaining black box models. J. ACM Comput. Surv. **51**(5), 1–42 (2018)
12. D. Gunning, Explainable artificial intelligence (XAI): technical report defense advanced research projects agency darpa-baa-16-53. J. DARPA, Arlington, USA (2016)
13. E. Tjoa, C. Guan, A survey on explainable artificial intelligence (XAI): Toward medical XAI. IEEE Trans. Neural Netw. Learn. Syst. (2020)
14. B. Goodman, S. Flaxman, European Union regulations on algorithmic decision-making and a "right to explanation". AI Mag **38**(3), 50–57 (2017)
15. W. McCullock, W. Pitts, A logical calculus of ideas immanent in nervous activity. Archive copy of 27 november 2007 on wayback machine. Avtomaty [Automated Devices] Moscow, Inostr. Lit. publ (1956), pp. 363–384
16. J. McCarthy, M. Minsky, N. Rochester, The dartmouth summer research project on artificial intelligence. Artif. Intell.: Past Present Future (1956)
17. A.M. Turing, Computing machinery and intelligence, in *Parsing the Turing Test* (Springer, 2009), pp. 23–65
18. S. Legg, M. Hutter et al., A collection of definitions of intelligence. Front. Artif. Intell. Appl. **157**, 17 (2007)
19. M. Newborn, *Kasparov Versus Deep Blue: Computer Chess Comes of Age* (Springer Science & Business Media, 2012)
20. T. Chouard, The go files: Ai computer wraps up 4-1 victory against human champion. Nat. News (2016)
21. A.C. Scott, R. Clancey, Explanation capabilities
22. A.C. Scott, W.J. Clancey, R. Davis, E.H. Shortliffe, *Explanation capabilities of production-based consultation systems* (Stanford University, CA, Deparment of Computer Science, Technical Report, 1977)
23. E.H. Shortliffe, A.C. Scott, M.B. Bischoff, A.B. Campbell, W. Van Melle, C.D. Jacobs, An expert system for oncology protocol management, in *Rule-Based Expert Systems*, ed. by B.G. Buchanan, E.H. Shortiffe (1984), pp. 653–655

24. M. Suwa, A.C. Scott, E.H. Shortliffe, An approach to verifying completeness and consistency in a rule-based expert system. AI Mag **3**(4), 16 (1982)
25. W.J. Clancey, The epistemology of a rule-based expert system—a framework for explanation. Artif. Intell. **20**(3), 215–251 (1983)
26. F. Xu, H. Uszkoreit, Y. Du, W. Fan, D. Zhao, J. Zhu, Explainable AI: a brief survey on history, research areas, approaches and challenges, in *CCF International Conference on Natural Language Processing and Chinese Computing* (Springer, 2019), pp. 563–574
27. P. Ongsulee, Artificial intelligence, machine learning and deep learning, in *2017 15th International Conference on ICT and Knowledge Engineering (ICT&KE)* (IEEE, 2017), pp. 1–6
28. J. Gerlings, A. Shollo, I. Constantiou, Reviewing the need for explainable artificial intelligence (XAI) (2020), arXiv preprint arXiv:2012.01007
29. M. Turek, Explainable artificial intelligence (XAI). DARPA (2018), https://www.darpa.mil/program/explainable-artificial-intelligence
30. D. Gunning, D. Aha, Darpa's explainable artificial intelligence (XAI) program. AI Mag. **40**(2), 44–58 (2019)
31. D. Gunning, M. Stefik, J. Choi, T. Miller, S. Stumpf, G.-Z. Yang, Xai—explainable artificial intelligence. Sci. Robot. **4**(37) (2019)
32. P. Voigt, A. Von dem Bussche, The EU general data protection regulation (GDPR), *A Practical Guide*, vol. 10, 1st edn. (Springer International Publishing, Cham, 2017), p. 3152676
33. P. Hacker, R. Krestel, S. Grundmann, F. Naumann, Explainable AI under contract and tort law: legal incentives and technical challenges. Artif. Intell. Law, 1–25 (2020)
34. A. Páez, The pragmatic turn in explainable artificial intelligence (XAI). Minds Mach. **29**(3), 441–459 (2019)
35. G. Vilone, L. Longo, Explainable artificial intelligence: a systematic review (2020), arXiv preprint arXiv:2006.00093
36. S. Hart, Shapley value, *Game Theory* (Springer, 1989), pp. 210–216
37. A. Henelius, K. Puolamäki, H. Boström, L. Asker, P. Papapetrou, A peek into the black box: exploring classifiers by randomization. Data Min. Knowl. Discov. **28**(5), 1503–1529 (2014)
38. E. Štrumbelj, I. Kononenko, Explaining prediction models and individual predictions with feature contributions. Knowl. Inform. Syst. **41**(3), 647–665 (2014)
39. E. Štrumbelj, I. Kononenko, M.R. Šikonja, Explaining instance classifications with interactions of subsets of feature values. Data Knowl. Eng. **68**(10), 886–904 (2009)
40. M.T. Ribeiro, S. Singh, C. Guestrin, Why should I trust you? Explaining the predictions of any classifier, in *Proceedings of the 22nd ACM SIGKDD International Conference on Knowledge Discovery and Data Mining* (2016), pp. 1135–1144
41. S.M. Lundberg, S.-I. Lee, A unified approach to interpreting model predictions, in *Proceedings of the 31st International Conference on Neural Information Processing Systems* (2017), pp. 4768–4777
42. J.H. Friedman, Greedy function approximation: a gradient boosting machine. Ann. Stat. 1189–1232 (2001)
43. A. Goldstein, A. Kapelner, J. Bleich, E. Pitkin, Peeking inside the black box: visualizing statistical learning with plots of individual conditional expectation. J. Comput. Graph. Stat. **24**(1), 44–65 (2015)
44. D.W. Apley, J. Zhu, Visualizing the effects of predictor variables in black box supervised learning models. J. Royal Stat. Soc.: Ser. B (Stat. Methodol.) **82**(4), 1059–1086 (2020)
45. B.M. Greenwell, B.C. Boehmke, A.J. McCarthy, A simple and effective model-based variable importance measure (2018), arXiv preprint arXiv:1805.04755
46. G. Hooker, Generalized functional Anova diagnostics for high-dimensional functions of dependent variables. J. Comput. Graph. Stat. **16**(3), 709–732 (2007)
47. A. Fisher, C. Rudin, F. Dominici, All models are wrong, but many are useful: Learning a variable's importance by studying an entire class of prediction models simultaneously. J. Mach. Learn. Res. **20**(177), 1–81 (2019)
48. M.T. Ribeiro, S. Singh, C. Guestrin, Anchors: high-precision model-agnostic explanations, in *Proceedings of the AAAI Conference on Artificial Intelligence*, vol. 32, no. 1 (2018)

49. C. Olah, A. Satyanarayan, I. Johnson, S. Carter, L. Schubert, K. Ye, A. Mordvintsev, The building blocks of interpretability. Distill **3**(3), e10 (2018)
50. D. Bau, B. Zhou, A. Khosla, A. Oliva, A. Torralba, Network dissection: quantifying interpretability of deep visual representations,' in *Proceedings of the IEEE Conference on Computer Vision and Pattern Recognition* (2017), pp. 6541–6549
51. K. Simonyan, A. Vedaldi, A. Zisserman, Deep inside convolutional networks: visualising image classification models and saliency maps (2013), arXiv preprint arXiv:1312.6034
52. G. Montavon, A. Binder, S. Lapuschkin, W. Samek, K.-R. Müller, Layer-wise relevance propagation: an overview, in *Explainable AI: Interpreting, Explaining and Visualizing Deep Learning* (2019), pp. 193–209
53. D.S. Yeung, I. Cloete, D. Shi, W.Y. Ng, Sensitivity analysis for neural networks (2010)
54. B. Kim, M. Wattenberg, J. Gilmer, C. Cai, J. Wexler, F. Viegas et al., Interpretability beyond feature attribution: quantitative testing with concept activation vectors (TCAV), in *International Conference on Machine Learning* (PMLR, 2018), pp. 2668–2677
55. Z. Chen, Y. Bei, C. Rudin, Concept whitening for interpretable image recognition. Nat. Mach. Intell. **2**(12), 772–782 (2020)
56. P.W. Koh, T. Nguyen, Y.S. Tang, S. Mussmann, E. Pierson, B. Kim, P. Liang, Concept bottleneck models, in *International Conference on Machine Learning* (PMLR, 2020), pp. 5338–5348
57. Q. Zhang, Y. . Wu, S.-C. Zhu, Interpretable convolutional neural networks, in *Proceedings of the IEEE Conference on Computer Vision and Pattern Recognition* (2018), pp. 8827–8836
58. A.B. Arrieta, N. Díaz-Rodríguez, J. Del Ser, A. Bennetot, S. Tabik, A. Barbado, S. García, S. Gil-López, D. Molina, R. Benjamins et al., Explainable artificial intelligence (XAI): concepts, taxonomies, opportunities and challenges toward responsible AI. Inform. Fus. **58**, 82–115 (2020)
59. C. Molnar, *Interpretable Machine Learning* (2019), https://christophm.github.io/interpretable-ml-book/
60. S. Lapuschkin, A. Binder, G. Montavon, K.-R. Muller, W. Samek, Analyzing classifiers: fisher vectors and deep neural networks, in *Proceedings of the IEEE Conference on Computer Vision and Pattern Recognition* (2016), pp. 2912–2920
61. S. Lapuschkin, A. Binder, G. Montavon, K.-R. Müller, W. Samek, The LRP toolbox for artificial neural networks. J. Mach. Learn. Res. **17**(1), 3938–3942 (2016)
62. S. Haneuse, D. Arterburn, M. . Daniels, Assessing missing data assumptions in EHR-based studies: a complex and underappreciated task. JAMA Netw. Open **4**(2), e210184–e210184 (2021)
63. R. Li, Y. Chen, J.H. Moore, Integration of genetic and clinical information to improve imputation of data missing from electronic health records. J. Am. Med. Inform. Assoc. **26**(10), 1056–1063 (2019)
64. B.K. Beaulieu-Jones, J.H. Moore, P.R.O.-A. A. C.T. Consortium, Missing data imputation in the electronic health record using deeply learned autoencoders, in *Pacific Symposium on Biocomputing 2017* (World Scientific, 2017), pp. 207–218
65. M.K. Hasan, M.A. Alam, D. Das, E. Hossain, M. Hasan, Diabetes prediction using ensembling of different machine learning classifiers. IEEE Access **8**, 76516–76531 (2020)
66. S.I. Ayon, M. Islam et al., Diabetes prediction: a deep learning approach. Int. J. Inform. Eng. Electr. Bus. **11**(2) (2019)
67. M.M. Mottalib, M.M. Rahman, M.T. Habib, F. Ahmed, Detection of the onset of diabetes mellitus by Bayesian classifier based medical expert system. Trans. Mach. Learn. Artif. Intell. **4**(4), 01 (2016)
68. G. Litjens, C.I. Sánchez, N. Timofeeva, M. Hermsen, I. Nagtegaal, I. Kovacs, C. Hulsbergen-Van De Kaa, P. Bult, B. Van Ginneken, J. Van Der Laak, Deep learning as a tool for increased accuracy and efficiency of histopathological diagnosis. Sci. Rep. **6**(1), 1–11 (2016)
69. M.S. Manak, J.S. Varsanik, B.J. Hogan, M.J. Whitfield, W.R. Su, N. Joshi, N. Steinke, A. Min, D. Berger, R.J. Saphirstein et al., Live-cell phenotypic-biomarker microfluidic assay for the risk stratification of cancer patients via machine learning. Nat. Biomed. Eng. **2**(10), 761–772 (2018)

70. N. Whiffin, R. Walsh, R. Govind, M. Edwards, M. Ahmad, X. Zhang, U. Tayal, R. Buchan, W. Midwinter, A.E. Wilk et al., Cardioclassifier: disease-and gene-specific computational decision support for clinical genome interpretation. Genet. Med. **20**(10), 1246–1254 (2018)
71. V. Fortino, L. Wisgrill, P. Werner, S. Suomela, N. Linder, E. Jalonen, A. Suomalainen, V. Marwah, M. Kero, M. Pesonen et al., Machine-learning–driven biomarker discovery for the discrimination between allergic and irritant contact dermatitis. Proc. Nat. Acad. Sci. **117**(52), 33474–33485 (2020)
72. M. Wu, J. Luo, Wearable technology applications in healthcare: a literature review. Online J. Nurs. Inform. **23**(3) (2019)
73. S. O'Sullivan, N. Nevejans, C. Allen, A. Blyth, S. Leonard, U. Pagallo, K. Holzinger, A. Holzinger, M.I. Sajid, H. Ashrafian, Legal, regulatory, and ethical frameworks for development of standards in artificial intelligence (AI) and autonomous robotic surgery. Int. J. Med. Robot. Comput. Assist. Surg. **15**(1), e1968 (2019)
74. A.-D. Dana, A. Alashqur, Using decision tree classification to assist in the prediction of Alzheimer's disease, in, *6th International Conference on Computer Science and Information Technology (CSIT)*, vol. 2014 (IEEE 2014), pp. 122–126
75. R.D. Gibbons, F.V. deGruy, Without wasting a word: extreme improvements in efficiency and accuracy using computerized adaptive testing for mental health disorders (CAT-MH). Curr. Psychiat. Rep. **21**(8), 1–9 (2019)
76. A. Suresh, R. Udendhran, M. Balamurgan, Hybridized neural network and decision tree based classifier for prognostic decision making in breast cancers. Soft Comput. **24**(11), 7947–7953 (2020)
77. S. Khare, D. Gupta, Association rule analysis in cardiovascular disease, in *2016 Second International Conference on Cognitive Computing and Information Processing (CCIP)* (IEEE, 2016), pp. 1–6
78. S. Agrawal, N. Mishra, Question classification for health care domain using rule based approach, in *International Conference on Innovative Data Communication Technologies and Application* (Springer, 2019), pp. 410–419
79. G. Wang, Z. Deng, K.-S. Choi, Detection of epilepsy with electroencephalogram using rule-based classifiers. Neurocomputing **228**, 283–290 (2017)
80. H. Byeon, Developing a random forest classifier for predicting the depression and managing the health of caregivers supporting patients with alzheimer's disease. Technol. Health Care **27**(5), 531–544 (2019)
81. C. Iwendi, A.K. Bashir, A. Peshkar, R. Sujatha, J.M. Chatterjee, S. Pasupuleti, R. Mishra, S. Pillai, O. Jo, Covid-19 patient health prediction using boosted random forest algorithm. Front. Pub. health **8**, 357 (2020)
82. P. Kaur, R. Kumar, M. Kumar, A healthcare monitoring system using random forest and internet of things (IoT). Multimed. Tools Appl. **78**(14), 19905–19916 (2019)
83. S.M. Mohnen, A.H. Rotteveel, G. Doornbos, J.J. Polder, Healthcare expenditure prediction with neighbourhood variables-a random forest model. Stat. Polit. Policy **11**(2), 111–138 (2020)
84. M. Dastoorpoor, N. Khanjani, A. Moradgholi, R. Sarizadeh, M. Cheraghi, F. Estebsari, Prenatal exposure to ambient air pollution and adverse pregnancy outcomes in Ahvaz, Iran: a generalized additive model. Int. Arch. Occupat. Environ. Health **94**(2), 309–324 (2021)
85. J. Yang, M. Zhang, Y. Chen, L. Ma, R. Yadikaer, Y. Lu, P. Lou, Y. Pu, R. Xiang, B. Rui, A study on the relationship between air pollution and pulmonary tuberculosis based on the general additive model in Wulumuqi, China. Int. J. Infectious Diseases **96**, 42–47 (2020)
86. R. Caruana, Y. Lou, J. Gehrke, P. Koch, M. Sturm, N. Elhadad, Intelligible models for healthcare: Predicting pneumonia risk and hospital 30-day readmission, in *Proceedings of the 21th ACM SIGKDD International Conference on Knowledge Discovery and Data Mining* (2015), pp. 1721–1730
87. L. Sagaon-Teyssier, A. Vilotitch, M. Mora, G. Maradan, V. Guagliardo, M. Suzan-Monti, R. Dray-Spira, B. Spire, A generalized additive model to disentangle age and diagnosis-specific cohort effects in psychological and behavioral outcomes in people living with hiv: the french cross-sectional anrs-vespa2 survey. BMC Pub. Health **19**(1), 1–10 (2019)

88. R. Yang, Who dies from Covid-19? post-hoc explanations of mortality prediction models using coalitional game theory, surrogate trees, and partial dependence plots. medRxiv (2020)
89. V. Gupta, M. Demirer, M. Bigelow, M.Y. Sarah, S.Y. Joseph, L.M. Prevedello, R.D. White, B.S. Erdal, Using transfer learning and class activation maps supporting detection and localization of femoral fractures on anteroposterior radiographs, in *IEEE 17th International Symposium on Biomedical Imaging (ISBI)*, vol. 2020 (IEEE 2020), pp. 1526–1529
90. S.D. Goodfellow, D. Shubin, R.W. Greer, S. Nagaraj, C. McLean, W. Dixon, A.J. Goodwin, A. Assadi, A. Jegatheeswaran, P.C. Laussen, Rhythm classification of 12-lead ECGS using deep neural networks and class-activation maps for improved explainability, in *Computing in Cardiology*, vol. 2020 (IEEE, 2020), pp. 1–4
91. S. Pereira, R. Meier, V. Alves, M. Reyes, C.A. Silva, Automatic brain tumor grading from mri data using convolutional neural networks and quality assessment, inUnderstanding and Interpreting Machine Learning in Medical Image Computing Applications (Springer, 2018), pp. 106–114
92. J. Irvin, P. Rajpurkar, M. Ko, Y. Yu, S. Ciurea-Ilcus, C. Chute, H. Marklund, B. Haghgoo, R. Ball, K. Shpanskaya et al., Chexpert: a large chest radiograph dataset with uncertainty labels and expert comparison. Proc. AAAI Conf. Artif. Intell. **33**(01), 590–597 (2019)
93. A. Kumar, S.B. Singh, S.C. Satapathy, M. Rout, Mosquito-net: a deep learning based CADX system for malaria diagnosis along with model interpretation using Gradcam and class activation maps. Exp. Syst. e12695 (2021)
94. M. Izadyyazdanabadi, E. Belykh, C. Cavallo, X. Zhao, S. Gandhi, L.B. Moreira, J. Eschbacher, P. Nakaji, M.C. Preul, Y. Yang, Weakly-supervised learning-based feature localization for confocal laser endomicroscopy glioma images, in *International Conference on Medical Image Computing and Computer-Assisted Intervention* (Springer, 2018), pp. 300–308
95. Y. Yang, V. Tresp, M. Wunderle, P.A. Fasching, Explaining therapy predictions with layer-wise relevance propagation in neural networks, in *2018 IEEE International Conference on Healthcare Informatics (ICHI)* (IEEE, 2018), pp. 152–162
96. G. Chlebus, N. Abolmaali, A. Schenk, H. Meine, Relevance analysis of MRI sequences for automatic liver tumor segmentation (2019), arXiv preprint arXiv:1907.11773
97. I.P. De Sousa, M.M.B.R. Vellasco, E.C. Da Silva, Local interpretable model-agnostic explanations for classification of lymph node metastases. Sensors (Basel, Switzerland) **19**(13) (2019)
98. S. Kitamura, K. Takahashi, Y. Sang, K. Fukushima, K. Tsuji, J. Wada, Deep learning could diagnose diabetic nephropathy with renal pathological immunofluorescent images. Diagnostics **10**(7), 466 (2020)
99. D. Sharma, A. Durand, M.-A. Legault, L.-P.L. Perreault, A. Lemaçon, M.-P. Dubé, J. Pineau, Deep interpretability for GWAS (2020), arXiv preprint arXiv:2007.01516
100. P.-Y. Tseng, Y.-T. Chen, C.-H. Wang, K.-M. Chiu, Y.-S. Peng, S.-P. Hsu, K.-L. Chen, C.-Y. Yang, O.K.-S. Lee, Prediction of the development of acute kidney injury following cardiac surgery by machine learning. Crit. Care **24**(1), 1–13 (2020)
101. T. Pianpanit, S. Lolak, P. Sawangjai, A. Ditthapron, P. Leelaarporn, S. Marukatat, E. Chuang-suwanich, T. Wilaiprasitporn, Neural network interpretation of the Parkinson's disease diagnosis from spect imaging (2019), arXiv e-prints, pp. arXiv–1908
102. A. Borjali, A.F. Chen, O.. Muratoglu, M.A. Morid, K.M. Varadarajan, Deep learning in orthopedics: how do we build trust in the machine? Healthc. Transform. (2020)
103. M.R. Zafar, N.M. Khan, Dlime: a deterministic local interpretable model-agnostic explanations approach for computer-aided diagnosis systems (2019), arXiv preprint arXiv:1906.10263
104. D. Yu, Z. Liu, C. Su, Y. Han, X. Duan, R. Zhang, X. Liu, Y. Yang, S. Xu, Copy number variation in plasma as a tool for lung cancer prediction using extreme gradient boosting (xgboost) classifier. Thoracic Cancer **11**(1), 95–102 (2020)
105. V. Couteaux, O. Nempont, G. Pizaine, I. Bloch, Towards interpretability of segmentation networks by analyzing deepdreams, in *Interpretability of Machine Intelligence in Medical Image Computing and Multimodal Learning for Clinical Decision Support* (Springer, 2019), pp. 56–63

106. S. Kandel, A. Paepcke, J. Hellerstein, J. Heer, Wrangler: interactive visual specification of data transformation scripts, in *Proceedings of the SIGCHI Conference on Human Factors in Computing Systems* (2011) pp. 3363–3372

107. K. Ham, Openrefine (version 2.5). http://openrefine.org, free, open-source tool for cleaning and transforming data. J. Med. Libr. Assoc.: JMLA **101**(3), 233 (2013)

108. D. Smilkov, N. Thorat, C. Nicholson, E. Reif, F. B. Viégas, M. Wattenberg, Embedding projector: interactive visualization and interpretation of embeddings (2016), arXiv preprint arXiv:1611.05469

109. D. Qiu, Y. Qian, Relevance debugging and explaining at linkedin (2019)

110. R. Florez-Lopez, J.M. Ramon-Jeronimo, Enhancing accuracy and interpretability of ensemble strategies in credit risk assessment. A correlated-adjusted decision forest proposal. Exp. Syst. Appli. **42**(13), 5737–5753 (2015)

111. A. Volkov, D.F. Benoit, D. Van den Poel, Incorporating sequential information in bankruptcy prediction with predictors based on Markov for discrimination. Decis. Supp. Syst. **98**, 59–68 (2017)

112. P. Xu, Z. Ding, M. Pan, An improved credit card users default prediction model based on ripper' in *2017 13th International Conference on Natural Computation, Fuzzy Systems and Knowledge Discovery (ICNC-FSKD)* (IEEE, 2017), pp. 1785–1789

113. P.M. Addo, D. Guegan, B. Hassani, Credit risk analysis using machine and deep learning models. Risks **6**(2), 38 (2018)

114. B. D. Radhakrishnan, J.J. Jaurez, Explainable artificial intelligence (XAI) in project management curriculum: Exploration and application to time, cost, and risk, in *2021 ASEE Virtual Annual Conference Content Access* (2021)

115. S. Afzal, A. Chaudhary, N. Gupta, H. Patel, C. Spina, D. Wang, Data-debugging through interactive visual explanations, in *Pacific-Asia Conference on Knowledge Discovery and Data Mining* (Springer, 2021), pp. 133–142

116. D.L. Marino, C.S. Wickramasinghe, M. Manic, An adversarial approach for explainable AI in intrusion detection systems (2018)

117. S. Mane, D. Rao, Explaining network intrusion detection system using explainable AI framework (2021)

118. S.R. Islam, W. Eberle, S.K. Ghafoor, A. Siraj, M. Rogers, Domain knowledge aided explainable artificial intelligence for intrusion detection and response (2020)

119. D. Rao, S. Mane, Zero-shot learning approach to adaptive cybersecurity using explainable AI (2021)

120. A. Morichetta, P. Casas, M. Mellia, Explain-it, in *Proceedings of the 3rd ACM CoNEXT Workshop on Big DAta, Machine Learning and Artificial Intelligence for Data Communication Networks*, December 2019. https://doi.org/10.1145/3359992.3366639

121. P. Giudici, E. Raffinetti, Explainable AI methods in cyber risk management. Qual. Reliab. Eng. Int. (2021)

122. B. Mahbooba, M. Timilsina, R. Sahal, M. Serrano, Explainable artificial intelligence (XAI) to enhance trust management in intrusion detection systems using decision tree model. Complexity **2021** (2021)

123. E. Wallace, S. Feng, J. Boyd-Graber, Interpreting neural networks with nearest neighbors (2018), arXiv preprint arXiv:1809.02847

124. N. Poerner, B. Roth, H. Schütze, Evaluating neural network explanation methods using hybrid documents and morphological agreement (2018), arXiv preprint arXiv:1801.06422

125. Y. Jiang, N. Joshi, Y.-C. Chen, M. Bansal, Explore, propose, and assemble: an interpretable model for multi-hop reading comprehension (2019), arXiv preprint arXiv:1906.05210

126. R. Ghaeini, X.Z. Fern, P. Tadepalli, Interpreting recurrent and attention-based neural models: a case study on natural language inference (2018), arXiv preprint arXiv:1808.03894

127. Y. Dong, Z. Li, M. Rezagholizadeh, J.C.K. Cheung, Editnts: an neural programmer-interpreter model for sentence simplification through explicit editing (2019), arXiv preprint arXiv:1906.08104

128. A. Abujabal, R.S. Roy, M. Yahya, G. Weikum, Quint: interpretable question answering over knowledge bases, in *Proceedings of the 2017 Conference on Empirical Methods in Natural Language Processing: System Demonstrations* (2017), pp. 61–66

129. R. Pryzant, S. Basu, K. Sone, Interpretable neural architectures for attributing an ad's performance to its writing style, in *Proceedings of the 2018 EMNLP Workshop BlackboxNLP: Analyzing and Interpreting Neural Networks for NLP* (2018), pp. 125–135

130. N. Liu, X. Huang, J. Li, X. Hu, On interpretation of network embedding via taxonomy induction, in *Proceedings of the 24th ACM SIGKDD International Conference on Knowledge Discovery & Data Mining* (2018), pp. 1812–1820

131. N. Pröllochs, S. Feuerriegel, D. Neumann, Learning interpretable negation rules via weak supervision at document level: a reinforcement learning approach, in *Proceedings of the 2019 Conference of the North American Chapter of the Association for Computational Linguistics: Human Language Technologies*, vol. 1 (Association for Computational Linguistics, 2019), pp. 407–413

132. K. Weitz, D. Schiller, R. Schlagowski, T. Huber, E. André, Do you trust me? Increasing user-trust by integrating virtual agents in explainable AI interaction design, in *Proceedings of the 19th ACM International Conference on Intelligent Virtual Agents* (2019), pp. 7–9

133. J. Druce, M. Harradon, J. Tittle, Explainable artificial intelligence (XAI) for increasing user trust in deep reinforcement learning driven autonomous systems (2021). arXiv preprint arXiv:2106.03775

134. D. Branley-Bell, R. Whitworth, L. Coventry, User trust and understanding of explainable AI: exploring algorithm visualisations and user biases, in *International Conference on Human-Computer Interaction* (Springer, 2020), pp. 382–399

135. J. van der Waa, E. Nieuwburg, A. Cremers, M. Neerincx, Evaluating XAI: a comparison of rule-based and example-based explanations. Artif. Intell. **291**, 103404 (2021)

136. A. Rai, Explainable AI: From black box to glass box. J. Acad. Mark. Sci. **48**(1), 137–141 (2020)

137. O. Biran, C. Cotton, Explanation and justification in machine learning: a survey, in *IJCAI-17 Workshop on Explainable AI (XAI)*, vol. 8, no. 1 (2017), pp. 8–13

138. T. Ha, Y.J. Sah, Y. Park, S. Lee, Examining the effects of power status of an explainable artificial intelligence system on users' perceptions. Behav. Inform. Technol. 1–13 (2020)

139. D. Wang, Q. Yang, A. Abdul, B.Y. Lim, Designing theory-driven user-centric explainable AI, in *Proceedings of the 2019 CHI Conference on Human Factors in Computing Systems* (2019), pp. 1–15

140. A. Papenmeier, G. Englebienne, C. Seifert, How model accuracy and explanation fidelity influence user trust (2019), arXiv preprint arXiv:1907.12652

141. R. Dazeley, P. Vamplew, C. Foale, C. Young, S. Aryal, F. Cruz, Levels of explainable artificial intelligence for human-aligned conversational explanations. Artif. Intell. **299**, 103525 (2021)

Chapter 2
An Overview of Explainable Artificial Intelligence for Cyber Security

Hind Khoulimi, Mohamed Lahby, and Othman Benammar

Abstract The rapid development of the Internet, various form of host and network attack have emerged, to detect and recognize different categories of attack Intrusion Detection System (IDS) have deployed as a defensive tool to detect attacks. However IDS based on manual and traditional techniques as signatures of known attacks and deviation of normal activity have become obsolete in the field of cyber security. Recently Artificial Intelligent (AI) especially Machine Learning (ML) and Deep Learning (DL) techniques are applied in IDS to construct a model which can be able to detect variety of attacks in real time. This work aims to provide an overview of various type of IDS, AI especially their two branches ML and DL. We also explain the importance of their conjunction in cyber security. Furthermore the different public dataset and various metrics used to analyze, compared and evaluate a ML and DL techniques for Intrusion Detection has been presented. Finally a series of discussion showed how AI enforce the effectiveness of cyber security. At the end of this chapter the different challenges of application of AI in Cyber Security are explored.

Keywords Cyber security · Artificial Intelligence · Deep learning · Intrusion detection system · Attack · Signature detection · Anomaly detection · Dataset · Performance evaluation

H. Khoulimi (✉) · M. Lahby · O. Benammar
Higher Normal School, Applied Mathematics and Computing Laboratory, Casablanca, Morocco
e-mail: hind-khoulimi-etu@etu.univh2c.ma

M. Lahby
e-mail: mohamed.lahby@univh2c.ma

O. Benammar
e-mail: othman.benammar@univh2c.ma

1 Introduction

The Internet has become more and more present in our daily lives, with its social networks and services as sharing information. Organizations have tended to open up to the opportunities offered by the Internet to adapt to the challenges of users and customers. All this, is possible thanks to technological projects and the expansion of networks all over the world. But, with every good invention, comes something bad. After the invention of the internet, came cyber threats therefore, that when companies began taking steps to secure their networks and information against cyber threats the focus was naturally on external intruders.

The need for IT security organizations is increasing due to the diversity of attacks, which become a major issue for researchers and professionals in this field.

For this reason, IT security technologies have been deployed such as antivirus (installed on machines connected to the network to check if viruses may have spread), authentication (to control access to data and resources), cryptography (the science that uses mathematics to encrypt and decrypt data to guarantee its confidentiality), the use of Virtual Private Network (VPN) between different agencies of the same organization to ensure the transfer and encryption of data passing through a public network, as well as firewalls which have gained in importance in order to secure the various existing vulnerabilities in a system; it consider as the first family of security mechanisms (when a firewall receives a packet, it begins by checking its Access list previously configured by network security experts [1]).

Nowadays, the growth of the Internet of things IoT and enormous amounts of data generated by different sensors and devices, the huge volume, velocity, variety and veracity of data received for analysis by current monitoring tools which becomes insuitable of handling the big amounts of data and detecting different attacks. At that time, Intrusion Detection System (IDS) which belong to the second family of security mechanisms capable of surrounding attacks and detecting them; emerged as a new and effective approach to protect computer systems and networks.

Several review articles shows that IDS is becoming an essential element of IT security in different architecture as Cloud, Internet of Things (IoT) and Mobile ad hoc Networks (MANET) (Table 1).

The existing security tools detect the attackers in offline mode, which become obsolete methods, because attackers use sophisticated techniques to integrate the infrastructure [2].

IDS based on fixed rules facilitate to attackers which deploys intelligent techniques to get control of their target. So a cat-and-mouse game between IT security and attackers didn't stop. To resolve this problem, the application of Artificial Intelligent (AI) especially Machine Learning (ML) and Deep Learning (DL) techniques in IDS is necessary with objective of building an intelligent IDS in order to achieve a good detection rate, low false alarms and in case of detection, taking appropriate actions if necessary.

Until today, researchers face the problem of building reliable and effective IDS that can handle numerous attacks with changing patterns [4].

Table 1 The use of IDS in security of the different new architectures in literature

Article	IoT	CLOUD	MANET
[2]			✓
[3]		✓	
[4]		✓	
[5]		✓	
[6]		✓	
[7]		✓	
[8]		✓	
[9]		✓	
[10]		✓	
[11]		✓	
[12]		✓	
[13]		✓	
[14]		✓	
[15]		✓	
[16]	✓		
[17]	✓	✓	
[18]			✓

2 Cyber Security

Cyber security has two components: security of information systems and computer security which deals with protecting of physical computers, laptops, and other devices. Cyber security can encompass all of these activities by protecting the data stored in networks and computers, etc. In other words Cyber security protects everything that is digitally connected. This security must be based on a reference document called a security policy. It is a document which contains all the rules that are defined after an analysis of the needs in term of resources and networks to be protected.

Talking about security leads us to distinguish between attack and anomaly, the first one is a deviation from the normal. We differentiate between:

- Internal and external attacks: the first one, relate to the cyber security risks that stem from the inside of any organization to exploit the system or to cause damage. The second one relates to outsider attacks attempting to gain unauthorized access to the network of the targeted organization.
- Malicious attacker: searches for a personal vulnerability by detecting weaknesses in the network in order to exploit it. It does not target any particular structure, it can be dangerous, but it is still easily identifiable.
- Rational attacker: targets specific points of the network with a specific objective. It's identification can be very difficult.

- Active attacker: the attacker can handle different acts on the network during the attack (by modifying, destroying, etc.)
- Passive attacker: the attacker listens to network messages and waits for useful information to launch an attack.

So the second one (anomaly) can be defined as an attempt to avoid security checks. If the attack successfully achieve the target, the attacker will have unlimited access to the victim (whether it's a server, a machine, etc.) and can do whatever he wants. The anomalies are categorized into three groups [12]:

- Point anomaly: if a single piece of data has deviated from the normal compared to the other data.
- Contextual anomaly: if a single piece of data is considered abnormal in a specific context, let us take the example of the temperature which may be low in summer, which is not normal in its context, because it is hot in the summer.
- Collective anomaly: if a group of data is abnormal compared to other data. This data group will be defined as a collective anomaly

There exists a different types of network attacks [12]:

- Denial of service: also known as Deni of service (Dos), it is an attack that aims to make computer equipment offline by sending numerous requests until it is saturated. There is a distributed form of this attack (Distributed Dos).
- Remote to Local (R2L): an attacker obtains full control of the targeted host by guessing his password via automated scripts.
- Probe or scan: A scan attack used to collect information and weaknesses of a machine or network. It can be considered a target identification tool to gain useful information to launch another dangerous attack.
- User to Root (U2R): the attacker aims to gain unauthorized access to an administrative account and gain privileges for exploiting important resources.

The main objective of these (anomalies, attacks) is to weaken and compromise the integrity, confidentiality, or availability of a resource: [16] So the security of information systems is based on three fundamental properties which are:

- Integrity: the ability of a security system to prevent alteration of data by prohibited entities. A security policy identifies which are authorized to modify information, but also which entities have the authorization to modify this information.
- Confidentiality: being able to protect data and resources and prevent illegitimate people from taking unauthorized access and rights.
- Availability: the ability of a system to provide access to data for these authorized users. So a security policy defines what services should be provided.

The simplest to the most complex attacks usually go through the following steps (see Fig. 1).

- Target identification: collect information on the target.
- Scanning: the attacker explores the network by looking for vulnerabilities.
- Exploitation: the flaws detected on the target are exploited by the attacker.

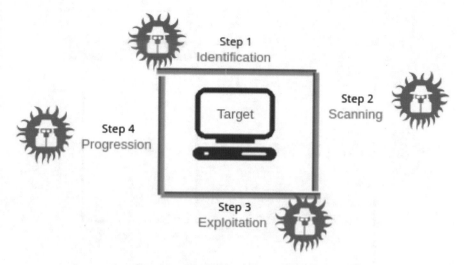

Fig. 1 Attack's steps

- Progress: the attacker takes control of the target (root), in order to be able to do whatever he wants (retrieving information, installing backdoors, cleaning up traces).

To reinforce the security mechanisms of detecting attacks, a second line of defense is deployed as a system of detection. IDS is a security detection system put in place to monitor networks or host (physical or virtual) for suspicious attacks [17] by observing actions, security logs or audit data using programs that collect, analyze and forwarded information to security administrators. It has been in existence since 1980s [5].

The raison of using an IDS is to analyze and detect any type of activity that must be compliant with security policy for detecting anomalies which cannot be detected by other security tools such as firewalls [18].

2.1   The Components of IDS

There are a common components of IDS (see Fig. 2):

- Traffic Collector: The component that is responsible for collecting data activities and event for analysis such as incoming and outgoing traffic, log and audit files and packets recorded by the operating system and network.
- Analysis Engine: Analyzes the data collected by the collector traffic. In the case of a knowledge-based IDS, the comparison of the data is performed with a signature database. On the other hand, if it is an IDS based on the behavior, it compares them with the normal behavior collected over time to see if the current behavior deviates or not.

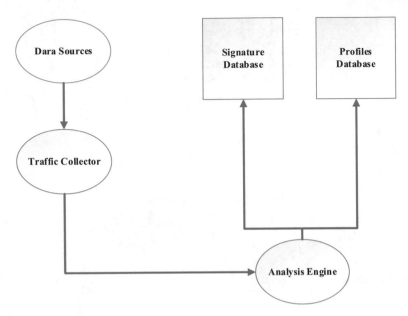

Fig. 2 The components of an intrusion detection system (IDS)

- Signature Database: Used in knowledge-based IDS, signature database is a fusion of signatures of attacks to be associated with suspicious and malicious activities.
- Management and Reporting Interface: allows system administrators to manage the IDS and receive alerts when intrusions are detected.

2.2 Intrusion Detection Process

In general, three essential components for IDS process (see Fig. 3) [19]:

- Data monitoring: Sensors are generally placed in specific places depending on what we want to observe (application, host or network) to collect and filter the relevant and necessary data to detect attacks as like (audit data, system data such

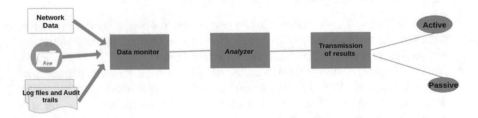

Fig. 3 General IDS process

as CPU load, number of active network connections, the number of repeated connection attempts, network data, source and destination address, etc.). In case of detecting attack alert will be send it to a secure console for analysis.

- Analysis: the analyzer proceed an analyze of information obtained in real time by choosing the most important ones in order to reduce the analysis time and make the analysis faster.
- The transmission of results: The IDS notifies the network administrator in case of detecting attack or suspicious system behavior.

There are several notification possibilities, the detection system can send an email with the nature of the attack, trigger a local alarm as a pop-up window in the security console and send an alert notification to a mobile's network administration.

After collecting, analyzing and generating an alarm, IDS must produce two different responses, the first one is the passive response of an IDS, in this case IDS record detected intrusions in a log file which will be analyzed by the security administrator in order to prevent recorded attacks from recurring in the future, but it does not directly prevent an attack to occurring. The second one is the active response which takes an automated actions when certain types of attacks are detected, such as stopping an attack by reconfiguring routers and firewalls to block attacker's packets, interrupting TCP connection (so as to prevent data transfer) and sometimes block the network in extreme situations.

As discussed in the thesis [1]: If the detection is in real time, the IDS can triggered an alert, which must be analyzed by the security administrator trying to take actions as blocking the network, but it can probably was notified after the intrusion ends. So the IDS must be coupled with a system that analyze alerts and take automated actions. For this reason combining the artificial intelligence with the human intelligence will be a good choice, because intrusion detection techniques alone will not prevent the attacks to happen.

The IDS can be classified according to the target monitoring, each of them presents a pros and cons (see Fig. 4):

- Network Intrusion Detection System (NIDS): are the popular types of IDSs, they rely on their own resources in terms of memory, CPU, etc. They can be placed at various locations on the network, upstream, or downstream of a firewalls, they analyze all the traffic entering and leaving the network to detect malicious activities such as Dos attacks, comparing to known attacks or behavior already observed. This sensor generally has two network cards, one placed in stealth mode on the network, the other allowing it to be connected to the security console [18].
- Host Intrusion Detection System (HIDS): Operating systems generally dispose an auditing system. The various system events are recorded in an audit log which should be analyzed frequently in order to collect information about the system and the behavior of its users.
- The log file formats are Common LogFile (CLF) and fExtended LogFormat (ELF), which are monitored by HIDS in order to analyze them and detect attacks. Noted that audit trails are more precise and detailed, whereas audit logs provide only essential information, but can be better analyzed due to their size.

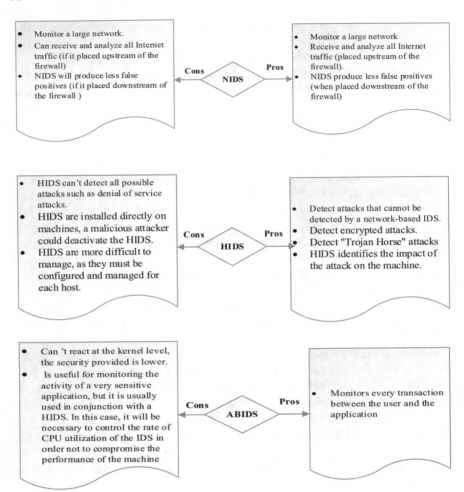

Fig. 4 The pros and cons of different types of IDS

- Application-based IDS (ABIDS): is a subgroup of HIDS, it controls the interaction between the user and an application program. It takes place at the level of communication between a user and the monitored application.
- Network Node Intrusion Detection System (NNIDS): Works like traditional NIDS which analyzes packets in network traffic. But NNIDS is only for packets destined for one address or a range of addresses.
- Stack-based IDS (SBIDS): working closely with the TCP/IP stack, monitored the packets moved up through the OSI layers which allows the IDS to remove the packets from the stack.

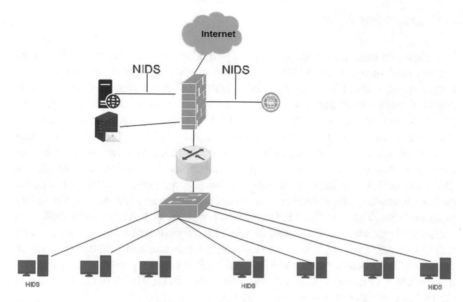

Fig. 5 The architecture of distributed detection systems

- Distributed IDS (DIDS): It is a distributed IDS, composed of several IDS such as (HIDS, NIDS) which communicates with each other and with a central server which brings together the information collected by the different HIDS and NIDS to analyze them and can detect known and unknown attacks seen that it benefits from both NIDS and HIDS (see Fig. 5).
- Hypervisor-based Intrusion Detection System (HBIDS): A hypervisor, also called a virtual machine monitor, is a process that creates and runs Virtual Machines (VMs). It allows a host to support multiple supported guest VMs by virtually sharing its resources, such as memory and processing capacity.
- The hypervisor-based intrusion detection system allows the user to monitor and analyze communications between VMs and between the hypervisor and VM and even their virtual network.
- Intrusion Prevention System (IPS): Monitors network traffic, and system activity to detect possible intrusions. It can block traffic and even with a false positive may block legitimate traffic. This is why a precise configuration must be deployed. We can distinguish between (HIPS, NIPS, WIPS, KIPS depending on the target to monitor either the host (VM, hypervisor), the network, the wireless network and even the core of the system).
- Intrusion Detection and Prevention System (IDPS): IDS and IPS are not able to provide full and effective security. So a combination between IDS and IPS remains necessary, which is called IDPS.

Generally, the most widely used in IDS are NIDS and HIDS in hybridization, ABIDS remain limited to use for extremely sensitive application.

2.3   Detection Mode

There are two main principals approaches to analyzing events to detect attacks the first one is **Misuse Detection (MD)** also known as abuse detection, malware recognition, Signature Based Intrusion Detection System (SIDS). Relies on libraries which contains attack description called signatures collected during network flow analysis. The IDS analyze each event and an alert will be triggered upon detects an attack resemble to signature that exists in the signatures database. There are two main techniques which are treated by [17] a Knowledge Based Approaches and Machine learning techniques (ML). The second famous approaches is **Anomaly Detection** (AD) also called Anomaly-based Intrusion Detection System (AIDS). AD operate on the assumption that the attack's activities are different from normal and legitimate activities. The behavioral approach was proposed by J. Anderson in 1980, then revised and extended by D. Denning in 1987, identify abnormal unusual behavior by calculating the rate of deviation between current behavior and past behavior which collected from historical data and statistical analyzes over a period of normal activities on a host or network. If this rate exceeds a certain threshold, the IDS release an alarm that computer system under attack [1].

The behavioral approach is based on several techniques such as observation of thresholds, user profiling, group profiling (to reduce the number of profiles to manage, users are classified into groups), resource use profiling, profiling of executable programs (consists of observing the use of system resources by certain executable programs in order to detect deviations from these behaviors for example, we can detect the fact that a server starts listening on ports other than those it usually uses), statistical profiling.

We can noted that exist a special approaches called **Specification-based Detection** is a set of rules that define the behavior of a network, namely nodes, protocols and routing tables. In the event of deviation from the rules, an intrusion is detected. We notice that has the same logic as anomaly detection approach identifies the deviations from the normal. However, the difference consist that rules must be manually entered [20]. Each of this mode has their pros and cons (see Fig. 6).

We can see in [17] the other new techniques emerged after the lunch of IDS virtualization.

To overcome the problems aforementioned before a combination of hybrid techniques is desired because the signature approach cannot detect an unknown attack, so the two detection mode (signature and anomaly) can work together in a single hybrid model. Any attack detected by the detection of an anomaly is transmitted to a signature database [21].

A especially in a cloud environment, hybridization is an inevitable need of the modes of detection mentioned above [17]. To reduce the rate of false positives generated by signature intrusion detection systems, and to have the ability to detect unknown attacks in the signature database.

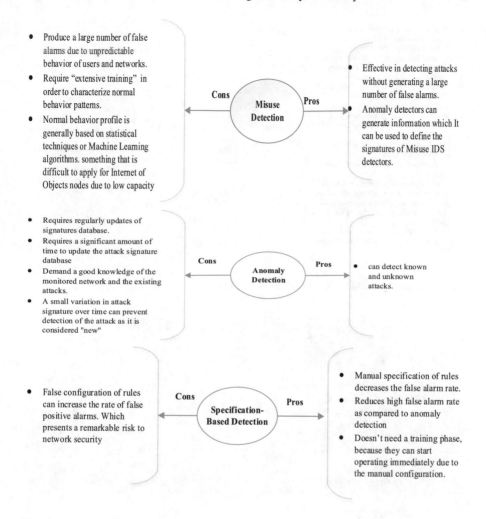

- Produce a large number of false alarms due to unpredictable behavior of users and networks.
- Require "extensive training" in order to characterize normal behavior patterns.
- Normal behavior profile is generally based on statistical techniques or Machine Learning algorithms. something that is difficult to apply for Internet of Objects nodes due to low capacity

Cons — Misuse Detection — Pros

- Effective in detecting attacks without generating a large number of false alarms.
- Anomaly detectors can generate information which It can be used to define the signatures of Misuse IDS detectors.

- Requires regularly updates of signatures database.
- Requires a significant amount of time to update the attack signature database
- Demand a good knowledge of the monitored network and the existing attacks.
- A small variation in attack signature over time can prevent detection of the attack as it is considered "new"

Cons — Anomaly Detection — Pros

- can detect known and unknown attacks.

- False configuration of rules can increase the rate of false positive alarms. Which presents a remarkable risk to network security

Cons — Specification-Based Detection — Pros

- Manual specification of rules decreases the false alarm rate.
- Reduces high false alarm rate as compared to anomaly detection
- Doesn't need a training phase, because they can start operating immediately due to the manual configuration.

Fig. 6 The pros and cons of detection mode

2.4 Strengths and Limitations of IDS

Intrusion Detection Systems are a valuable addition to an organization's security infrastructure, but like any IT mechanism has strengths and weaknesses (Table). Many factors affect the efficiency of an IDS solution such as its positioning (which depends on the network topology) or its sensors, its mode of operation, its configuration which depends on the requirements of the network and its services in order to provide an effective intrusion detection solution (see Fig. 7).

The effectiveness of an IDS depends on several criteria such as the detection mode (signature, anomaly or hybrid), the location of the IDS (host, network, virtual

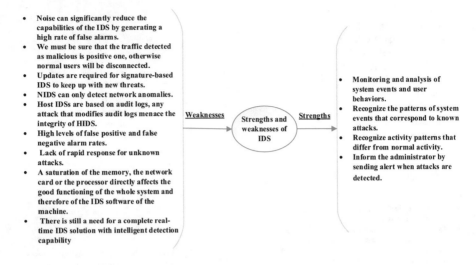

Fig. 7 The strengths and weaknesses of IDS

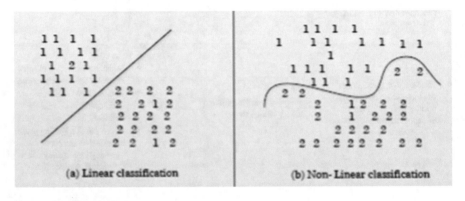

Fig. 8 Classification [27]

machine, hypervisor or distributed) and its configuration. Researchers face the problem of building reliable and effective IDS that can handle numerous attacks with changing patterns [4].

Indeed, the complexity of information systems, the quantity and variation of the data collected, the enormous number of connected objects which continues to increase have made these statistical mechanisms incapable of processing all this data [22].

Our study will be focused on the use of intelligent methods and techniques, which are learned from their experiences automatically without prior configuration to help IDS for detecting attacks. This is called Artificial Intelligence (AI) with its two branches Machine Learning (ML) and Deep Learning (DL). These ML and DL techniques have been used in computer network security in order to analyze data to identify anomalies. These techniques are generally three types: supervised learning, unsupervised learning and semi-supervised learning. Each technique is based on two phases training and testing phase.

3   Artificial Intelligence

AI is a field that is currently presented in several fields "e.g." health, banking, marketing and sales of a company, computer network as well as robots and autonomous cars these are forms of AI. The goal is to make them smarter [23].

Our study will focus on the use of smarter methods and techniques, which are learned from their experiences automatically without prior configuration. Using AI with its two branches ML and DL.

These techniques generally work on three stages [24]:

- **Preprocessing**: this is a step that consists of properly structuring the data in order to prepare them for the training phase.
- **Training**: this is a step that makes it possible to extract useful information from the data prepared prior to the preprocessing phase, in order to build a data model.
- **Detection**: based on the model developed by the previous phase, the supervised traffic data will be the input of the model which will compare them with the one already created. So that it can classify them into two classes normal or not.

And what is worse is that the attackers themselves exploit these ML techniques in order to overcome the lines of defense of the security system. Something that proves that traditional computer security practices are no longer effective security of the line of defense remains a major issue and the best solution is to use ML and DL.

- Machine Learning

ML is a subdomain of AI [25]. Is the field of study that gives computers the ability to learn and improve from experiences without being explicitly programmed. It has been applied to automatic automobiles, marketing, networking, chemistry, medicine, pharmacy or energy.

Deep Learning

DL or Deep Neural Networks (DNN) is a subdomain of ML, it is based on the use of artificial neurons inspired by the human brain. These neurons are organized in layers thus giving the notion of depth. This is where the term deep comes from. Are much more complex models that represent mathematical functions with a large number of parameters. Something that requires a larger dataset, a longer learning time and a lot of computing power.

- Learning Methods

The concept of learning is generally based on two stages treated successively:

- Memorization: learning and memorizing the different cases encountered.
- Generalization: the fact of being able, to deal with distinct examples, not yet encountered, but similar.

So, the questions should ask are: How does ML and DL works? How does the machine learn? The machine relies on learning techniques that are inspired by the same way humans learn things.

- Supervised learning

These techniques are generally three types:

Supervised learning is the most used in ML, which provides both the input data (input) and the expected output data (output), during a phase called training the input and output data are labeled for classification, in order to establish a learning base for further processing of the data [26].

In the case of an IDS, the collected traffic is classified into two main classes (normal, abnormal). Even that can go a long way, classify it as a different class of attack type. So we use supervised learning when it is necessary to converge towards a precise final state.

The developed models of supervised aims to solve two types of problems:

- **Regression problems**: when trying to predict the value of a continuous variable which can have an infinity of values.
- **Classification problems**: as its name indicates, the aim is to classify a discrete object which can have only a limited number of values (Fig. 8).

- Unsupervised learning

Supervised learning consists of learning to the machine things that we already known, we build in advance a Dataset which contains X questions and Y answers. If we take the case of a network of the IoT type where objects are connected to each other via the Internet where their capacities in terms of memory and deployed computer resources are limited, something which restricts the use of more sophisticated models to detect unknown attacks. Something that proves that in this case unsupervised learning remains the best choice [28]. So if the Dataset does not have Y values. Unsupervised techniques do not rely on any training phase and learn directly from unlabeled data [26]. During unsupervised learning, the system converge to any end state, by looking for associations between attributes without using classes or labels. The case of anomaly detection, normal network behavior is more common and which aggregates into a denser cluster which indicates, while malicious behavior is rarer and which appears as outliers and more distant in the entire Dataset [29].

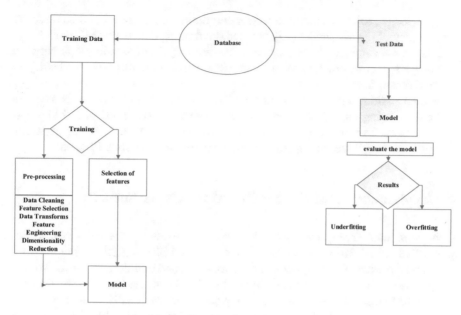

Fig. 9 The process of training an ML and DL model

- Semi-supervised learning

Semi-supervised learning falls between unsupervised learning and supervised learning, when there is not enough labeled data to produce an accurate model, semi-supervised learning techniques can be used to augment the size of the training data.[1]

Each technique is based on two phases namely:

- The training phase
- The test phase

The general process of learning is as follows (see Fig. 9):

- **Data collection**: either in offline mode when based on data history, or online using network protocol monitoring tools, namely SNMP (Simple Network Management Protocol).
- **Data preparation (Feature engineering)**: this is the phase of cleaning the data before its use in order to ensure that they are complete, non-redundant, without unnecessary entries and without noise, based on the Principal Component Analysis method (PCA). The aim is to extract useful information that will serve the next phase. Then the selection of targeted data for evaluation in order to extract

---

[1] https://medium.com/cuelogic-technologies/evaluation-of-machine-learning-algorithms-for-int rusion-detection-system last accessed 2020/06/25.

knowledge [30]. The goal is to reduce computational overhead and increase the accuracy of the ML model [29], we feed the training data to our ML model.

- **Model creation**: using ML techniques, an analysis of relationships between complex data in order to extract knowledge and learn to properly classify or predict new data.
- **Model validation**: The evaluation of the model's performance. This step can be done using one of two methods: cross validation (holdout) or k-fold cross-validation. The last step, evaluates the ML model on the basis of accuracy, precision, recall receiver operating characteristic (ROC) curve, F1 score.

4   Explainable Artificial Intelligent for Cyber Security

Cyber-criminals deploy evolving attack patterns to evade the detectors. Traditional IDS (NIDS/HIDS) especially signature-based NIDS/HIDS, require continuous updates. However the growing of network data and traditional tools for ML are lacking as we move towards the distributed and real time processing [30]. So integrating the hybrid ML algorithms for the big data processing will help to increase the percentage of producing accuracy in results and also reveal the hidden knowledge of the big data.

Traditional NIDS suffer from high false recognition rate which can be reduced with ML and DL techniques compared to manually or statistical techniques. So the detection rate will be improved with reduced of False Negatives.

Therefore, AI especially ML and DL techniques have been actively studied to help the IDS to resist malicious attacks based on the assumption that the patterns of the attack packets differ from those of the normal packets [31]. Based on a set of features that identify a state of an object which should be chosen carefully and accurately to avoid inaccuracy and unnecessary time consumption. To achieve good data learning with accurate and efficient results, unrelated or irrelevant features must be removed from the set. Such a process is known as features reduction or dimensionality reduction. It is necessary to choose and generate only important features which called feature selection.

Dash [32] Clarify that in recent studies techniques such as artificial neural network (ANN), Decision Tree (DT), Evolutionary Algorithms (EA), Genetic Algorithm (GA), Particle Swarm Optimization (PSO), Simulated Annealing (SA)), Rule-based Data Mining, Swarm Intelligence have gained significance attention to solve the problem of intrusion detection.

In the literature different NIDS and HIDS was deployed with ML/DL techniques. So it will difficult to comparing them because each one treat special attacks, in a special architecture, with different dataset, using various ML and DL algorithms.

As like [33] proposed a Feed-forward Neural Network (FNN) combined with Locust Swarm Optimization (LSO) (FNN-LSO) to build an advanced detection system and improve the performance of IDS. Their results confirm that LSO is recommended only when the datasets and the number of features are very large. The LSO algorithm is useful for large datasets because of the extreme number of local optima that make the conventional training algorithm almost in effective.

Dash [32] has been combined the Gravitational Search (GS) and PSO algorithms to train a neural network for different real-world datasets and have proved to be successful in its applications to illustrate the applicability of the proposed approaches, the performance measures (e.g., error, training time, decision time, overage detection, accuracy rate) have been obtained by testing with NSL-KDD dataset, and the results have been compared with popularly used such as Decision Tree (DT), Gradient Descent (GD-ANN), Genetic Algorithm (GA-ANN) and Particle Swarm Optimization (PSO-ANN).

4.1 The Performance Evaluation and Pubic Dataset

To compare the effectiveness of ML and DL techniques to classify an unknown instance, a set of criteria and indicators (Table 2) such as accuracy, precision, training time [25] (Table 3) this criteria differ depending on the problem to be solved and the objectives to be achieved.

There are applications that require a high detection rate with fast response, others need reliable detection instead of fast response. In order to evaluate the performance of intrusion detection systems, and ML/DL techniques, a set of databases are used (Table 2). The choice of data to test and evaluate the model remains a challenge given that several reasons to know [34].

- Data inconsistency.
- Redundancy and duplication of features.
- Missing values.

KDD 99 is the most used reference, NSL-KDD 99 is the filtered version of KDD99 where the number of redundancies is reduced and minimized. Thakkar and Lohiya [35] contains normal attributes and four types of attacks namely (DoS, U2R, R2L and Probing) [35].

To evaluate the performance of an ML and DL model (see Fig. 10), we calculate the difference between the actual values and the expected values of ML or DL model for classification or regression problems. The most used metric to evaluate the classification model is what is called "ACCURACY". The evaluation of an ML or DL model used in intrusion detection (IDS) is generally based on two parameters: the detection rate and the false alarm rate, in order to identify the accuracy of the system at the detection level. The ideal is to have an IDS system that provides high detection and less false alarms which is difficult to achieve. As research shows that increasing the detection rate itself increases the false alarm rate [23]. The parameters used to evaluate the performance of ML or DL model in conjunction with an intrusion detection system are [36, 37].

Table 2 Public datasets

Datasets	Description
KDD Cup'99	Was generated by The Massachusetts Institute of Technology (MIT) sponsored by DARPA (US Defense Research and Development Agency) in 1998 (DARPA 1998). But with the rapid development of technology and the emergence of new attacks, a new version of this database appeared in 1999 (DARPA 1999). This database contains 4 types of attacks: DoS, U2R, Probe and R2L
NSL-KDD	The NSL_KDD database was created in 2009 to solve some KDD cup 99 problems such as the removal of redundant connections that degrade the performance of learning algorithm
UNSW-NB15	Created by the Australian Center for Cyber Security (ACCS) which contains two million records. This dataset contains attacks types such as (Worms, Shellcode, Reconnaissance, Port Scans, Generic, Backdoor, DoS, Exploits and Fuzzers)
CIC-IDS 2017	Created by the Canadian Institute for Cyber security (CIC) in 2017, which contains normal flows and attacks from the real world. Such as (DDoS, Botnet, DoS, web attack….)
CSE-CIC-IDS 2018	Created by Telecommunications Security Establishment (CSE) and Canadian Institute for Cybersecurity (CIC) in 2018. It includes seven different attack scenarios: Brute-force, Heartbleed, Botnet, DoS, DDoS, Web attacks, and infiltration of the network from inside
ADFA-LD 12	Created by (Australian Force Academy), it was designed for anomaly-based IDS to evaluate an HIDS system
ICS X 2012	Developed by the Center of Excellence for Information Security at the University of New Brunswick. it contains seven-day captured traffic with a total of 2,450,324 streams, including DoS attacks
TON-IoT	Proposed by (ALSAEDI et al. 2020) in 2020 collected from operating system logs, IoT network traffic it has been used to evaluate several ML and DL techniques

Table 3 Confusion matrix

Predicted outcome		
Actual result	Negative	Positive
	True negative	False positive
	False negative	True positive

- TP: True Positive (a real attack with a triggered alarm)
- FP: False Positive (incorrect rejection)
- TN: True Negative (identified as normal)
- FN: False Negative (equivalent to those incorrectly admitted)
- Accuracy: The ratio of correct classifications

$$\text{Accuracy} = \frac{TP + TN}{TP + TN + FP + FN}$$

Fig. 10 General Structure of
evaluation of ML/DL model

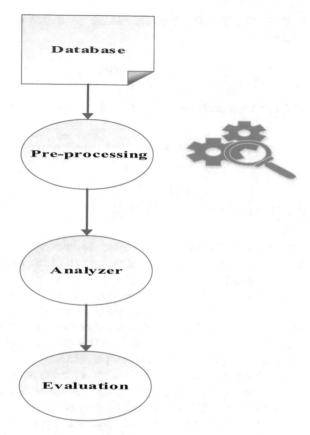

- Precision: (The detection rate) the number of correct classifications versus the
  number of missed entries

$$\text{Precision} = \frac{\text{TP}}{\text{TP} + \text{FP}}$$

- Recall: (detection rate) the number of correct classifications versus the number of
  missed entries

$$\text{Recall} = \frac{\text{TP}}{\text{TP} + \text{FN}}$$

- FPR: False Positives Rate is the proportion of negatives incorrectly detected as
  positives.

$$\text{FPR} = \frac{\text{FP}}{\text{FP} + \text{TN}}$$

- TPR: The True Positives Rate is the proportion of positives that are correctly detected.

$$TPR = \frac{TP}{TP + FN}$$

- F1 Score is the harmonic mean of the precision and the TVP.

$$F1 \text{ score } = \frac{2 \times \text{ precision } \times PR}{\text{Precision } + TPR}$$

5 Discussion

Nowadays, Cyber security is became the important field due to the development technologies on the one hand and the growth of connectivity devices via Internet in the other hand.

However, it is important to note that to understand how to interpret ML and DL algorithms in Cyber security, we first need to understand what a ML and DL algorithms which are different to traditional programming where pass several examples and let ML and DL learn whet the best (lowest cost) decisions are instead of writing every action that the algorithm must perform.

In other words, ML and DL algorithms receive inputs are able to transform them into outputs without being explicitly programmable. The importance of the using AI with two branches ML and DL techniques in computer security which is demonstrated and approved by various articles. Each ML and DL techniques tested on one of the databases already discussed (Table) to strengthen Cyber security mechanism could achieve different performance metrics by changing the database and techniques deployed which makes the choice of suitable technique for a specific problem remain a major issue.

However, deploying the techniques of classification, clustering, statistical, information theoretic, spectral anomaly detection, and nearest-based anomaly detection to improve detection process. Comparing their results we can notice that anomaly detection with labeled output called supervised techniques are more effective than unsupervised techniques, the biggest advantages consist of recognizing known malicious activities with high accuracy and low false alarm rate, the only weaknesses is that they have a less ability to recognize 0-days attacks. Also techniques based on clustering, information theory and nearest neighborhood are better compared to classification and statistical techniques. Even that classification and clustering are more effective to identify DoS attacks compared with other techniques. Also several experiences have shown that K-means clustering is an effective choice to grouping data into three classes (3?), and Naïve Bayes classifier can classify data into five classes (Normal, DoS, Probe, R2L and U2R).

The Nearest Neighbor and Clustering techniques are not effective when the number of dimension is high, so the differentiation between a normal and attack instance becomes more difficult.

Support Vector Machine (SVM), Artificial Neural Network (ANN), Fuzzy Association, Decision Tree and Statistical models are the most techniques applied in IDS for Cyber security.

With the big amounts of data even ML techniques becomes incapable to handle the data that's why researchers in Cyber security have started to apply DL which become important and applicable for intrusion detection as its achieves better performance comparing to traditional ML techniques for IDS, several studies have approved this as like An IDS based on DL algorithms using Recurrent Neural Network (RNN), the experiment which is on NSL-KDD dataset has shown high accuracy compared to ML techniques with only one drawback consist of pending more time on training that's why the authors proposed using GPU acceleration which can reduce training time.

However combining Deep Neural Network with Association analysis technology to classifying, detecting and generating abnormal traffic alarms with an important reduction in false alarms.

Some techniques which doesn't require a training phase as Nearest Neighbor, information theoretic and spectral techniques requires a significant time of testing. Therefore, classification, clustering and statistics techniques which require more training time, but the test is more rapid which is acceptable as long as the training is done in offline mode, but the test is done in real time. We note that reducing classifier training time may improve IDS performance.

Researchers in Cyber security have testing ML and DL techniques on IDS by several experimentation, we remarked that the famous datasets is NSLKDD which not contains all attacks, which makes testing the performance of ML and DL models uncertain especially in the case of IoT which is considered a new type of network architecture. However this dataset contain a redundant data which can decrease the performance of IDS. An important step in constructing ML and DL models is the selection and extraction of useful features from the traffic monitored, this features can change their values during an attack such as source address, destination address, packet source, etc. That's why the authors trying to choosing the useful features because selecting wrong features which changes values during an attack even the best algorithm will never be able to detect an intrusion because is a most important step that plays an important role in reaching a significant accuracy and decreasing FP and FN.

Generally, deploying an IDS capable of detecting attacks in real time is difficult because detection requires a prior knowledge of characteristics of the attacks, which remain impossible.

To resolve many of weaknesses of deploying a ML or Dl techniques only, the authors are directed to a hybridization between ML and Dl algorithms in order to identify multiple types of attack and improve classifier performance.

Recently, attacker targets ML and DL models in order to damage the training process to prevent the model making a correct classification or prediction. That's why we need a system which assures the protection of the training phase and the confidentiality as well as the integrity of the data. As declared in his article currently we need new ML algorithms adopted to computer security since ML, DL and Cyber security are two different fields. So expertise in these two fields is necessary, because in the beginning of ML techniques are not designed to detect attacks. So this combination between ML/DL and IDS is new; this explains the non-performance and all the weaknesses already mentioned before. So it is a new voice for research in this field.

Several review papers [12, 22, 25, 38, 39] demonstrate the importance of ML and DL techniques in computer security, but each of these techniques has strengths and limitations, which become more difficult to choose the appropriate technique used for a particular problem, as the same ML and DL techniques could achieve different performances by changing the case. So, choosing the good technique for a specific problem remains a major issue.

The author [12] worked on network anomaly detection techniques (classification, clustering, statistical, information theoretic), [40] as well works with other detection techniques (spectral anomaly detection, nearest- based anomaly detection).The two authors agrees on the following points: Anomaly detection techniques with labeled output are more effective than unsupervised techniques [12], techniques based on clustering, information theory and nearest neighborhood are better compared to classification and statistical techniques [12, 40], the techniques based on classification and clustering are interesting to identify DoS attacks which confirms by [41] the authors find that K-means clustering is very good for grouping data into three classes and Naive Bayes classifier to classify data into five classes (Normal, DOS, Probe, R2L and U2R) [41].

Support Vector Machine (SVM), Artificial Neural Network (ANN), Fuzzy Association, decision tree and statistical models are the techniques most used in IDS [42]. As deep learning has become important and applicable for intrusion detection, several studies have shown that deep learning achieves better results than traditional methods for intrusion detection [43], an intrusion detection system based on a deep learning approach using RNN (Recurrent Neural Network), the experiment which is based on the NSL-KDD dataset, has shown high accuracy compared to traditional machine learning methods with only one drawback consist of spending more time on training even the authors proposed using GPU acceleration can reduce training time.

Researchers in Computer Security have started to apply deep learning techniques as it achieves better objectives than traditional ML techniques [37] which proves the authors themselves combining (DNN 4 & Association analysis technology) to classify, detect and generate abnormal traffic alarms with a remarkable reduction in false alarms.

Some techniques which doesn't require a training phase like (nearest neighbor, information theoretic and spectral techniques), require a significant time at the test level [40]. On the other hand, classification, clustering and statistics are techniques which require significant training time, however the test is more rapid, which is acceptable as long as the training is done in offline mode, but the test is done in real time. We note that reducing classifier training time may improve IDS performance [44].

Chandola et al. [40] Confirms that nearest neighbor technique and clustering techniques when the number of dimension is high at that time, the differentiation between a normal and attack instance becomes more difficult.

The articles treated shows that the best known supervised techniques are Support Vector Machine, Bayesian Network, Artificial Neural Network, Decision Tree and k-Nearest Neighbor. The biggest advantage is that they recognize well-known malicious activity with high accuracy and low false alarm rate. The only weaknesses is that they have a less ability to recognize 0-Day attacks.

By reviewing the most experiments of application of the ML/DL techniques on IDS [48–51, 53–55, 57–61, 63], we remarked that the famous datasets utilized by the authors is NSLKDD which lacked of all available attacks, which makes testing the performance of ML and DL models in certain specially in the case of Internet of Things (IoT) which is considered to be a new type of network architecture and new traffic.

This dataset have a redundant data which are used to evaluate the performance of IDS can decrease the performance of IDSs [58].

An important step in forming ML and DL models is the selection and extraction of useful features from the traffic [59] this features can change values during an attack such as (source address, destination address, packet source, destination source) by comparing them with a normal period. The authors confirms we must makes attention to choosing the useful features because if we choose features which doesn't makes any change values during an attack then even the best algorithm will never be able to detect an intrusion. This is a most important step that plays an important role in reaching a significant accuracy and decreasing FP and FN [60].

Generally, deploying a IDS capable of detecting attacks in real time is almost impossible task because detection requires prior knowledge of the characteristics of the attacks, which confirms by [61].

Dixit and Silakari [22] Confirmed the need for hybridization between ML and DL algorithms in order to identify multiple types of attack and improve classifier performance as [44] Proposed a model that uses SVM and extreme learning to improve detection of known and unknown attacks, [62] proposed a model that uses SVM and extreme learning to improve detection of known and unknown attacks, [63] proposed a model based on Support Vector Machine (SVM) and K-nearest neighbor (KNN), [64] Proposed a Cloud-IDS which captures packets from the network, analyzes them based on K-Nearest Neighbor and Neural Network (KNN-NN) and generates a report. This model also combines detection by signature and detection by profile.

Recently, attackers targets ML and DL models in order to damage the training process to prevent the model to making a correct classification or prediction [58]. That's why we will need a system that ensures the protection of the training phase and the confidentiality as well as the integrity of the data, as declared [29] in his article currently we need new ML algorithms adopted to computer security since ML, DL and computer security are two different fields So expertise in these two fields is necessary. Which confirmed as well [43]. In the beginning ML techniques are not designed to detect attacks. So this combination between ML/DL and IDS is new; this explains the non-performance and all the weaknesses already mentioned before. So it is a new voice for research in this field [28].

6 Addressed Challenges for AI in Cyber Security

IDSs are considered a second line of defense which have shown performance in terms of attack detection, but still suffer from limitations that must be resolved in order to achieve a high and optimal level of security.

- Many models of ML/DL suffers of Overfitting when a model fits exactly against its training data. When this happens, the algorithm unfortunately cannot perform accurately against unseen data. So these models when implemented in a real environment do not show accuracy. However when the model memorizes the noise and fits too closely to the training set, the model becomes "overfitted" and it is unable to generalize well to new data. If a model cannot generalize well to new data, then it will not be able to perform the classification or prediction tasks that it was intended for. So to resolve this challenge, the data has to be labeled, something which is difficult to do, and takes a long time, or reduce complexity in the model by eliminating less relevant inputs. However, if you pause too early or exclude too many important features, you may encounter the opposite problem, and instead, you may underfit your model. Underfitting occurs when the model has not trained for enough time or the input variables are not significant enough to determine a meaningful relationship between the input and output variables. So the best solution is DL as it learns by extracting what is useful from the data without the need for manual labeling [28].
- However, the key research question is how we can find a model that avoids both these two problems? For this, there is a method that allows to limit the variance of a model without sacrificing its bias is regularization, an efficient technique to detect if the model has a bias (Under fitting) or variance (Over fitting) problem which consists in analyzing the errors (the Cost Function) on the Train Set and the Test Set. If the errors are large on the Train Set and the Test Set, then the model has a large bias which we can remedy them by creating a more complex model, with more parameters, we can even train the model more long time. If the errors are small on the Train Set, but are large on the Test Set, then the model has a large variance. So, the optimal bias-variance model is the one that manages to have the least possible bias while maintaining less variance [65].
- Most studies use the NSL-KDD and KDD-Cup 99 standard public databases for training ML/DL algorithms which are obsolete [37]. So a major challenge is building a database that contains almost all types of attacks, and which must be updated daily to contain the latest signatures of attacks in different network's architectures as IoT and of course make it public to help researchers testing the effectiveness of their algorithms in Cyber security [66].

- Even, researchers has been developed a models using supervised and unsupervised techniques, and they have got a high accuracy rate. Researchers cannot decide whether their models will detect new attacks due that this models was trained using obsolete dataset.
- To achieve high accuracy of ML, DL or ML-DL model, high CPU performance is required, and selection of the most important features to accelerate data pre-processing [66]. But a the big challenge is the nature of anomalies still continues to change and attackers adapt their network attacks to escape detection intrusion [40].
- In case of the Internet of Things (IoT), most of researchers are focused on the development of IDS to detect a few types of attacks, something that decreases the performance of IDS because it does not cover all types of attacks.
- Different Datasets are not applicable to test the validity of IDS intended for IOT due to the huge need for a database that shows the various attributes and attacks of an Internet of Things network.
- As well as IoT and mobile networks continue to grow and with the lack of suitable IDS conforms of the architecture of IoT which inquire a combination between hybrid IDS (signature and anomaly), ML/DL techniques and BigData to guarantee a good security [28].

7   Conclusion

The security of a network is an increasingly complex and extensive task as the network continue to grow and Cyber-attacks are increasing which overcoming traditional measures of security. However, Security presents a significant challenge for specialists and researchers in the field of protecting applications, machines and networks.

IDSs are essential tools for good network security considered as second line of defense after the firewalls, but this system based on statistical techniques present a limitation in detecting and preventing all types of attack's. The comparative study allowed us to discover intrusion detection systems, their types (application, host and network) as well as their detection methods (signature, anomaly, etc.), and present an overview of ML and DL techniques explaining how it works, focusing on explaining the significant role in controlling different attacks using ML and DL. For this reason, researchers have proposed several models based on ML and DL to improve detection techniques in order to achieve a high level of detection.

Due to the novelty of AI especially a ML and DL techniques in cyber security different challenges are discussed in this chapter. These challenges require further research to propose an intelligent system based on ML and DL techniques which can detect, analyze and take appropriate and suitable actions if necessary in order to achieve a good accuracy and precision with a low false alarm rate.

References

1. A. Kartit, Une nouvelle approche de détection d'intrusions et étude des problèmes liés au déploiement de politiques de sécurité dans les réseaux informatiques (2011)
2. J.-M. Percher, B. Jouga, Détection d'intrusions dans les réseaux Ad hoc, *Projet* (2003)
3. S. Roschke, F. Cheng, Meinel, Intrusion detection in the cloud (2009)
4. A. Bakshi, Y.B. Dujodwala, Securing cloud from DDOS attacks using intrusion detection system in virtual machine (2010)
5. S.N. Dhage, B.B. Meshram, R. Rawat, S. Padawe, M. Paingaokar, A. Misra, Intrusion detection system in cloud computing environment (2011)
6. T. Alharkan, P. Martin, IDSaaS: intrusion detection system as a service in public clouds (2012)
7. C. Modi, D. Patel, B. Borisaniya, H. Patel, A. Patel, M. Rajarajan, A survey of intrusion detection techniques in Cloud (2013)
8. M. Ficco, L. Tasquier, R. Aversa, Intrusion detection in cloud computing (2013)
9. A. Patel, M. Taghavi, K. Bakhtiyari, J. Celestino Ju´nior, An intrusion detection and prevention system in cloud computing: a systematic review (2013)
10. M. Jouini, L. Ben Arfa Rabai, Surveying and analyzing security problems in cloud computing environments (2014)
11. E. Vasilomanolakis, S. Karuppayah, M. Muhlh ¨ Auser, M. Fischer, Taxonomy and survey of collaborative intrusion detection (2015)
12. A. Mohiuddin, M. Abdun Naser, J. Hu, A survey of network anomaly detection techniques (2016)
13. S. Singh, Y.-S. Jeong, et J. Hyuk park, A survey on cloud computing security: issues, threats, and solutions (2016)
14. A. Chiba, N. Abghour, K. Moussaid, A. El omri, M. Rida, A cooperative and hybrid network intrusion detection framework in cloud computing based on snort and optimized back propagation neural network (2016)
15. M.S. Dildar, N. Khan, J. Bin Abdullah, A. Shahid Khan, Effective way to defend the hypervisor attacks in cloud computing (2017)
16. P. Parrend, J. Navarro, F. Guigou, A. Deruyver, P. Collet, Foundations and applications of artificial Intelligence for zero-day and multi-step attack detection (2018)
17. P. Mishra, E.S. Pilli, V. Varadharajan, U. Tupakula, Intrusion detection techniques in cloud environment: a survey (2017)
18. D. C. Modi, B. Borisaniya, H. Patel, A. Patel, M. Rajarajan, A survey of intrusion detection techniques in Cloud (2012)
19. R. Heenan, N. Moradpoor, A survey of intrusion detection system technologies (2016)
20. B.B. Zarpelão, M.R. Sanches, C.T. Kawakani, S.C. de Alvarenga, A survey of intrusion detection in internet of things (2017)
21. M. Elmubarak, A. Karrar, N. Hassan, Implementation Hybrid (NIDS) System using Anomaly Holtwinter Algorithm and signature based scheme (2019)
22. P. Dixit, S. Silakari, Deep learning algorithms for cybersecurity applications: a technological and status review (2020)
23. Q. Wang, P. Lu, Research on application of artificial intelligence in computer network technology (2018)
24. M.G. da Silva Neto, D.G. Gomes, Network intrusion detection systems design: a machine learning approach (2017)
25. A.L. Buczak, E. Guven, A survey of data mining and machine learning methods for cyber security intrusion detection (2016)
26. D. Ucci, L. Aniello, R. Baldoni, Survey of machine learning techniques for malware analysis (2018)
27. N. Moustafa, J. Hu, J. Slay, A holistic review of network anomaly detection systems: a comprehensive survey (2018)
28. R. Ahmad, I. Alsmadi, Machine learning approaches to IoT security: a systematic literature review (2021)

29. R. Boutaba et al., A comprehensive survey on machine learning for networking: evolution, applications and research opportunities (2018)
30. R. Ahamed et al., Real-time big data processing for anomaly detection: a survey (2018)
31. M.-J. Kang, J.-W. Kang, Intrusion detection system using deep neural network for in-vehicle network security (2016)
32. T. Dash, A study on intrusion detection using neural networks trained with evolutionary algorithms (2015)
33. I. Benmessahel, K. Xie, M. Chellal, T. Semong, A new evolutionary neural networks based on intrusion detection systems using locust swarm optimization (2018)
34. K. Siddique, Z. Akhtar, F.A. Khan, Y. Kim, KDD cup 99 data sets: a perspective on the role of data sets in network intrusion detection research (2019)
35. A. Thakkar, R. Lohiya, A review of the advancement in intrusion detection datasets (2020)
36. B. Hu, J. Wang, Y. Zhu, T. Yang, Dynamic deep forest: an ensemble classification method for network intrusion detection (2019)
37. M. Gao, L. Ma, H. Liu, Z. Zhang, Z. Ning, J. Xu, Malicious network traffic detection based on deep neural networks and association analysis (2020)
38. D. Kwon, H. Kim, J. Kim, S.C. Suh, I. Kim, K.J. Kim, A survey of deep learning-based network anomaly detection (2017)
39. Z. Ahmad, A. Shahid Khan, C. Wai Shiang, J. Abdullah, F. Ahmad, Network intrusion detection system: A systematic study of machine learning and deep learning ap)proaches (2020.
40. V. Chandola, A. Banerjee, V. Kumar, Anomaly detection: a survey (2009)
41. M.H. Bhuyan, D.K. Bhattacharyya, J.K. Kalita, Network anomaly detection: methods, systems and tools (2014)
42. K. Shaukat et al., Performance comparison and current challenges of using machine learning techniques in cybersecurity (2020)
43. C. Yin, Y. Zhu, J. Fei, X. He, A deep learning approach for intrusion detection using recurrent neural networks (2017)
44. W. Laftah Al-Yaseen, Z. Ali Othman, Multi-level hybrid support vector machine and extreme learning machine based on modified K-means for intrusion detection system (2017)
45. C. Yin, A deep learning approach for intrusion detection using recurrent neural networks (2017)
46. K. Sydney Mambwe, A deep learning method with filter based feature engineering for wireless IDS (2019)
47. G. Wang, J. Hao, J. Ma, L. Huang, A new approach to intrusion detection using Artificial Neural Networks and fuzzy clustering. Expert Syst. Appl. 37(9), Article no 9 (2010). https://doi.org/10.1016/j.eswa.2010.02.102.
48. S. Mohammadi, Cyber intrusion detection by combined feature selection algorithm (2018)
49. A. Faour, P. Leray, C. Foll, Réseaux bayésiens pour le filtrage d'alarmes dans les systèmes de détection d'intrusions, in EGC (Ateliers) (2005), pp. 69–72
50. M. Gao, L. Ma, H. Liu, Z. Zhang, Z. Ning, J. Ma, Malicious network traffic detection based on deep neural networks and association analysis (2020)
51. M.G. da Silva Neto, Network intrusion detection systems design: a machine learning approach (2017)
52. W. Qingjun, L. Peng, Research on application of artificial intelligence in computer network technology (2018)
53. T. Geethapriya, S. Chawla, Towards deep-learning-driven intrusion detection for the internet of things (2019)
54. L. Yun, The individual identification method of wireless device based on dimensionality reduction and machine learning (2017)
55. N. Thanh Van, T.N. Thinh, L. Thanh Sach, A combination of temporal sequence learning and data description for anomaly based NIDS\Kopernio (2019). https://kopernio.com/viewer?doi=10.5121%2Fijnsa.2019.11307&token=WzE5OTA3MTQsIjEwLjUxMjEvaWpuc2EuM jAxOS4xMTMwNyJd.qACLFwWBHQahUczAeY2nVpAJAAI (consulté le avr. 24, 2020)

56. A. Javaid, Q. Niyaz, W. Sun, M. Alam, A deep learning approach for network intrusion detection system, in *Présenté à 9th EAI International Conference on Bio-inspired Information and Communications Technologies (formerly BIONETICS)*, New York City, United States (2016). https://doi.org/10.4108/eai.3-12-2015.2262516.
57. M.A. Ambusaidi, H. Xiangjian, N. Priyadarsi, T. Zhiyuan, Building an intrusion detection system using a filter-based feature selection algorithm (2016)
58. I. Sharafaldin, A. Habibi Lashkar, A.A. Ghorbani, Toward generating a new intrusion detection dataset and intrusion traffic characterization (2018)
59. M. Zolanvari, M.A. Teixeira, L. Gupta, K.M. Khan, R. Jain, Machine learning network vulnerability analysis of industrial (2019)
60. A. Amouri, V.T. Alaparthy, S.D. Morgera, A machine learning based intrusion detection system for mobile internet of things (2020)
61. N. Hoque, M.H. Bhuyan, R.C. Baishya, D.K. Bhattacharyya, J.K. Kalita, Network attacks: taxonomy, tools and systems (2014)
62. A.-Y. Wathiq Laftah, Multi-level hybrid support vector machine and extreme learning machine based on modified K-means for intrusion detection system (2017)
63. A. Abdulla Amin, B.I.R. Mamun, A novel SVM-kNN-PSO ensemble method for intrusion detection system (2016)
64. P. Ghosh, A.K. Mandal, R. Kumar, An efficient cloud network intrusion detection system, in *Information Systems Design and Intelligent Applications*, vol. 339, ed. by J.K. Mandal, S.C. Satapathy, M. Kumar Sanyal, P.P. Sarkar, A. Mukhopadhyay (Springer India, New Delhi, 2015), pp. 91–99. https://doi.org/10.1007/978-81-322-2250-7_10
65. V. Gustavsson, Machine learning for a network based intrusion detection system (2019)
66. P. Kumar, A.A. Kumar, C. Sahayakingsly, A. Udayakumar, Analysis of intrusion detection in cyber attacks using DEEP learning neural networks (2020)

Chapter 3
Artificial Intelligence: Practical and Ethical Challenges

Sophie Rice and Leandros Maglaras ⓘ

Abstract Artificial Intelligence (AI) & Machine Learning (ML) is used everywhere in daily life from speech recognition on our phones, targeted marketing strategies & face recognition without the majority of society even realising it. The power of technology and availability of data has exploded in recent years with new and exciting technological advancements being developed every single day. This chapter aims to highlight and critically evaluate the practical and ethical challenges brought about by the advancement of AI and ML as well as identifying any research gaps and suggesting areas where further exploration is required.

Keywords Artificial intelligence · Ethical design · Big data · Privacy

1 Introduction

The concept of Artificial Intelligence (AI) has been around far longer than one would have initially presumed, with some researchers suggesting the idea can date all the way back to the ancient civilisations engineers building automatons, and classical philosophers attempting to describe human thinking as a symbolic system. The most recognisable form however is much more recent, with the earliest notable workings emerging in the late 1950's where the term 'Artificial Intelligence' was first coined [18]. Although initially it was thought that AI held great promise for its future capability, its growth inhibited after several developmental failures and plateaus in new research in the field. In short, people lost interest. Scientists accepted it was an unfeasible feat and turned there attentions elsewhere. Fast-forward to today, AI & Machine Learning is used everywhere in daily life from speech recognition on

S. Rice (✉)
Technology—Cyber and Information Security, British Telecom, London, UK
e-mail: sophie.rice@bt.com

L. Maglaras
Cyber Technology Institute, De Montfort University, Leicester, UK
e-mail: leandros.maglaras@dmu.ac.uk

© The Author(s), under exclusive license to Springer Nature Switzerland AG 2022   59
M. Ahmed et al. (eds.), *Explainable Artificial Intelligence for Cyber Security*,
Studies in Computational Intelligence 1025,
https://doi.org/10.1007/978-3-030-96630-0_3

our phones, targeted marketing strategies & face recognition without the majority of society even realising [11].

The power of technology and availability of data has exploded in recent years with new and exciting technological advancements being developed every single day. For machine learning in particular, we are only just scratching the surface with the possibilities and opportunities it could bring for the enhancement of society with particular focus on the benefits of AI in Medicine and the improvements it can make within the healthcare field. As is expected though, all of this progress into new and unchartered territory does not come without a price. Machine learning relies on the availability and accessibility of the ever growing amount of data produced by the billions of interactions made by people every day. "The collection, storage, use & sharing of [this data] poses many ethical questions regarding governance, quality, safety, standards, privacy and data ownership" [29] as well as creating several practical challenges in regards to how to even process this vast amount of information with most of the frameworks currently available simply unable to cope. Even the use of trusted networks (e.g blockchain) cannot eliminate the security and privacy risks of information leak or misuse [9].

Research has found that the majority of available sources take the form of scholarly works, with in depth descriptions on the theory & mechanics behind machine learning, as well as a heavy focus in the healthcare sector and how machine learning can and is changing medical care. It is noted however, that nearly all researched sources were published in the past 3 years, emphasising the topics immaturity and suggesting there is a real need for further investigation. This chapter aims to highlight and critically evaluate the practical and ethical challenges brought about by the advancement of machine learning as well as identifying any research gaps and suggesting areas where further exploration is required.

The rest of this chapter is organized as follows: In Sect. 2, we present Machine Learning principles. In Sect. 3 we analyze the interplay between ML and Big Data. In Sects. 4 and 5 we present practical and ethical issues of AI. Finally, Sect. 6 includes the conclusions that we draw from this research.

2 Machine Learning

In a report published by The Royal Society, Artificial Intelligence is defined as "the science of making machine's smart". It's a branch of technology that allows machines to perform activities we commonly associate with human thinking such as learning, decision making & problem solving. Machine Learning is a subset of AI, a discipline that "allows computer systems to learn directly from examples, data and experience" [26]. Similarly, Géron defines Machine Learning as programs that are capable of learning from data without being explicitly programmed. It's the idea that machines can be autonomous, independently thinking for themselves without the need for constant human input or intervention. It has been suggested that the developments seen in ML today originated from a paper published by Geoffrey Hinton in [13] that detailed how to successfully train a deep neural network to recognise handwritten digits, a method that was to be branded 'deep learning'. A deep neural network is, in the simplest terms, a technological replica of the human brain. Its stacked layers of artificial neurones fire and communicate much like the biological neurones in our own brains would [11].

Machine Learning utilizes several types of learning techniques to 'train' a program or model, with the most dominant two being supervised learning which allows the reasonable prediction of future outputs based on known or labelled data inputs and responses and unsupervised learning which allows the finding of hidden patterns within unlabelled datasets; essentially with this technique the model is "learning without a teacher" [11]. These categories can be split into different tasks for which there is an associated algorithm. **Supervised learning** commonly uses classification or regression algorithms, whereas **unsupervised learning** uses clustering or association algorithms among others.

– An example of **supervised** learning: Classification Task
  Spam Filter—Our email spam filter identifies features of a new mail and classified whether or not it should be sent to the spam folder or our normal inbox.
– An example of **supervised** learning: Regression Task
  Numeric Value Prediction—The price of a property can be estimated given a set of features such as square footage, location, number of bedrooms etc. based on known data of previous house prices and features.
– An example of **unsupervised** learning: Clustering
  Running a clustering algorithm using data about a website's visitors to try and detect groups of similar visitors, finding connections without human assistance or data labelling.
– An example of **unsupervised** learning: Association
  Market Based Analysis—Used by retailers by analysing transactions to find associations between items to see which are frequently bought together in order to tailor recommendations.

There are many other categories, techniques and tasks associated to machine learning but the outputs are all ultimately determined in the same way; by the data fed into each algorithm called a 'training set'. Both Géron [11] and Rajkomar et al. [25] agree that in order to be successful, the training set needs to be large, representative and unpolluted with minimal noise. Essentially, you get out what you put in. If you put in poor data, the machine will output a poor result. Booch [5] describes the process of machine learning as "gaining enough knowledge to win". In theory, the model needs every possible example in order to build an extensive knowledge base and learn effectively from them.

3   Big Data and Machine Learning

Big Data is the term used to describe the process of collecting and analysing the enormous amount of data created as a result of the billions of online interactions made every day [7]. Online doesn't necessarily always mean via the internet, rather by any means where the information is stored or captured electronically. It is dynamic, ever evolving and theoretically unlimited with no one fixed source. This information can include public, open & personal data, and it can't be handled by traditional data processing methods as they are not designed to be able to cope with its unpredictable nature that needs to be dealt with in near real-time.

Research has shown that Big Data can be defined by 5 main characteristics;

1. Volume (Scale of Data)
   The amount of data to be managed, usually measured in Terabytes or Petabytes.
2. Velocity (Data in Motion)
   The frequency or speed of the data that is produced, processed & analysed
3. Veracity (Uncertainty of Data)
   Can also relate to quality of data, refers to uncertainty due to data inconsistency, incompleteness, ambiguities, latency & deception.
4. Variety (Different forms of Data)
   Big Data can come in all different formats e.g. structured numeric in databases or unstructured text documents, emails, audios, videos and more.
5. Value (Data's Worth)
   The worth of information to various stakeholders or decision makers [17].

Machine Learning uses the data provided by Big Data to fulfil the demands necessary for it to be able to perform successfully. With the huge amount of information at its disposal, Big Data can provide the machines with enough varied examples to learn from in order to produce better results and encourage a better performance from the ML algorithms [31]. Obviously, the enormous cost and effort required to execute these tasks effectively means the people behind making it work must be investing their time and resource for a reason, which leads us onto the question: what is its purpose?

Mendelson [20], Jones et al. [15] and Krishna et al. [17] all agree that the main purpose of Machine Learning is to improve Quality of Life. They provide an "opportunity for better decision making" allowing us to improve the manageability and accuracy of everyday tasks [2]. Some real life examples include voice recognition technologies such as Apple's Siri or Amazon's Alexa which enable hands free living, and ML medical technologies which can detect cancerous tumours. In short, machine learning can make life easier. It can automate repetitive tasks and solve problems in a fraction of the time a human brain could, allowing us to turn our attentions elsewhere. For businesses it could mean removing the manual labour and human error of number crunching, or for the already overstretched healthcare sector it could mean removing manual diagnosis time. Moreover, ML has been proven to be efficient in cybersecurity applications [8, 10, 16].

The question remains however, where do we draw the line? Research has shown that the average person is more accepting of AI when it's proven to be useful to benefit people in some way, but are understandably cautious when faced with its future capabilities [26]. Already we have seen Machine Learning technologies developed simply because we can, such as the computer that can beat the world's greatest chess player for no other reason than to prove it can be done. Jones et al. [15] raised this point by quoting Stephen Hawking in saying AI "could be the best or worst thing to happen to humanity". If this area of technology continues to grow unchecked with no legal boundaries or frameworks to work to it may end up inflicting more harm than good. A selection of the practical and ethical challenges we should be mindful of as this field progresses are explored below.

4 Practical Challenges of Machine Learning

In this section we present the main practical challenges that arise from the use of Machine Learning techniques.

4.1 Sourcing the Data

As we have previously established, in order to be successful ML models require a huge amount of diverse data to create representative training sets to train the machines. Without this data the machine would only output unscalable, unrepresentative and unusable results. With this in mind, we can assume that it is a massive challenge for companies or corporations who are attempting Machine Learning to source this data.

Both Rajkomar [25] and Krishna [17] have identified obtaining high quality data as one of the main challenges associated with machine learning, alongside being able to process data meaningfully. This would suggest that merely acquiring any form of available data is not enough, it has to be purposeful and relevant. Big Data certainly plays a hand in providing the backbone of ML data however it is not without restrictions. Certain sensitive data is protected by legal regulations such as GDPR and companies face huge fines if found in breach of the rules. For example, Facebook and Google have both paid huge sums in fines for violating the terms and conditions of data handling, mostly for transporting it across country lines to different database locations [20]. Some 3rd parties may choose to buy data from other companies however similar restrictions still apply [24].

In essence, data for use in machine learning must be acquired legally, ethically and if applicable with consent while still remaining representative and maintaining quality. Even for the technology giants this is not an easy task and comes alongside the additional prerequisite of storing this data.

4.2   Data Storage and Processing

Research from the IDC has shown that the amount of data that exists in the world is doubling every two years. Its rate of growth is outpacing the rate in which we can effectively interpret it and is also proving too much for the existing standard data handling models [23]. As such, the key players in the Machine Learning have turned to cloud based solutions to both store and process their data in real-time. Although the cloud offers the lure of essentially unlimited storage capabilities, it comes at a considerable cost not just financial. Both Hesamifard et al. [12] and Mendelson [20] put forward the argument that the introduction of Machine Learning as a Service (MLaaS) brings increased risk.

In most cases, the entire dataset would have to be sent over to the 3rd party cloud provider. This gives them visibility and also access to potentially sensitive data, leaving its confidentiality and integrity vulnerable to compromise [30]. Extra precautions would need to be taken to ensure the security of the data both in transit to the cloud and at rest with the 3rd party provider to prevent any security breaches. It is not just in MLaaS where there is risk however, as the very nature of the ML model leaves training set data exposed to malicious activity. It's easy to extract sensitive data from the model parameters, and any change in the data by a malicious end user is very easily missed meaning fake data could be influencing the machines output decisions [20]. Essentially, Machine Learning models in their current state are nowhere near as secure as they should be and this is an area that requires a significant amount of resource and testing. There are ways to patch over the problem areas with methods such as data obfuscation for the datasets and access control for who can view and manage the data however with the capabilities of malicious threat actors evolving every day it is essential that more is done.

Other practical issues related to big data and ML are [6]:

– The classic needle-in-a-haystack problem
– Data silos
– Inaccurate data
– Lack of skilled workers
– Future security issues

Currently, the power of Machine Learning is only available to the large corporations such as Google, Amazon & Microsoft who can afford the huge cost in resource it requires to operate. In theory, the capabilities of Machine Learning are endless yet it must be remembered that there are numerous critical dependencies for its success that are exceptionally difficult to meet, even more the most well-established & well-funded developers.

5   Ethical Challenges of Machine Learning

In this section we present the main ethical challenges that arise from the use of Machine Learning techniques.

5.1   *Data Privacy and Security*

In recent years after a number of high profile data breaches there has been a significant spotlight shining on data security. After the introduction of GDPR in 2016 the general public are becoming more security conscious and there is increased pressure on those in control of our data to ensure they are handling it correctly. For Machine Learning specifically though, both Panch et al. [22] and Mendelson [20] argue that GDRP and other similar regulations aren't enough. There are significant gaps they do not cover, and there is a real issue of non-conformity for example between US and UK regulations. Rajkomar et al. [25] supports this argument by highlighting the gap between what legally binds a person or a group of people when handling data and the legal boundaries of a machine. For example, It would be illegal for a physician to have unlimited access to a patients entire medical history & personal information yet this is standard practice for the Machine Learning models used in medical diagnosis. There is an inherent lack of legal frameworks explicitly exclusive to Machine Learning and its use of data which leads us to believe that although it may be handled legally it is likely not adhering to ethical considerations.

There is an expectation for there to be a certain level of transparency when it comes to what happens to our data. We should be able to know how it is being used to determine the decisions made by a machine however some ML models, in particular deep learning machines, use black box algorithms which display the input and output data yet obfuscate the internal decision making process making it illegible to the human end user. This breaches the GDPR "right to explanation" clause, where a person affected by the decision made has the right to request an explanation for how the particular decision was made [22].

Additionally, it is always assumed that the use of any data requires consent yet there is an often overlooked grey area around data used in Machine Learning and how consent is obtained. Often, a data privacy statement is presented to anyone inputting personal data somewhere. It is likely that the majority of people will tick the 'I agree' box without reading the small print, essentially giving 3rd parties full freedom to use their data outside of the initial purpose without realising while the 3rd party can argue they have given 'informed consent'. On the other hand, fully informed consent may be given initially however we have to take into account that consent may change over time. A research example shows an experiment spanning 18 years that documented patient medical data with consent only being established at the beginning [17]. This data went on to be published to a publicly available source which could easily be re-used in machine learning models without explicit consent, which suggests that more legal focus needs to be placed on consent for use and re-use of personal data [27].

Furthermore, there could be an instance where sensitive personal data can be used in a Machine Learning model without the need for consent. Mendelson [20] provides discussion around the social credit system being developed by the Chinese government which utilizes a form of machine learning that would track the movements of every citizen using data collected from street CCTV cameras & reports from banks, schools, department stores and so on. The purpose is to reward or punish citizens depending on their social-credit score decided by a black box ML algorithm using the collected data. This type of system gives a small amount of people absolute control over the people and their data which violates not only data privacy but privacy as a whole. Being a government program there is little in the way of legal restrictions and it is unlikely the citizens will be given a choice about their data being used in this way which once again proves what is legal is not always ethical. Mendelson also suggests there are similar programs to this already in development which leads us to believe this type of data use will start to become a lot more commonplace in the near future, especially since we cannot currently control or prevent the intentional/unintentional collection of our personal data.

5.2   Bias and Fairness

There is an overwhelming amount of research that points to unconscious bias being one of the main challenges associated with machine learning that is also arguably the most difficult to overcome [1, 4, 14, 15, 28]. There are two ways in which an ML model could display bias;

1. Algorithms inherit and repeat subjective bias found in historical data used in its training sets.
2. The model correctly finds a particular attribute of individuals is valuable in predicting outcomes, where society may deem use of such an attribute as inappropriate.

Bird et al. [4] suggests that the legal framework that prohibits the use of attributes such as race or gender in decision making tasks significantly reduces the risk of bias however The Royal Society report [26] argues that Machine Learning actually destabilises the distinction between sensitive & non-sensitive data, allowing datasets to be used in ways that allow sensitive to be derived from the mundane. For example, a user's Facebook likes may be used to predict attributes such as gender, age, sexuality and political stance which can then influence the decision making process. Veale and Binns [28]) even go as far as to argue that Machine Learning models are designed to discriminate and that access to personal attributes in fact helps to avoid bias as the model requires knowledge of protected characteristics to be successful, inferring that the legal framework restrictions actually encourage bias instead of eliminating it. However, their research also suggests that "individuals are unlikely to be happy providing a comprehensive range of personal data to the very organisations who are in a position to discriminate, no matter how technically robust their anonymization process is" [28]. This highlights the difficulties faced by companies looking to implement machine learning in the attempt to balance unfairness, data privacy and social expectations.

The difficulty with overcoming unconscious bias is that it is for the most part, unconscious, yet will have a real impact on individuals and groups even helping to perpetuate bias in some cases. It will likely only be identified after either an analysis on the training data or on the outputted decisions specifically checking for any evidence of unfairness. With no concrete and realistically feasible method to completely eliminate unconscious bias, it is evident this needs to remain at the forefront of ongoing research to find a resolution as Machine Learning continues to mature.

Amazon screening Algorithm. In 2018, Amazon trialled a new AI recruiting tool they had developed that used machine learning to search the web for potential candidates, giving them a rating from 1 to 5 stars. In this case, the training dataset was formed from previously submitted CV's from a 10 year period which gave examples of how to spot the best candidates.

However, Amazon decided to shut the tool down after the algorithm was found to discriminate against women, learning to systematically downgrade women's CV's for more technical jobs such as engineering roles or software development. Despite several attempts, Amazon could not find a way to make its algorithm gender-neutral, despite being at the forefront of AI technology.

There are several possible reasons for the failure. Firstly, with the low proportion of women working at the company especially in these sorts of roles, the machine identified male dominance as a factor in success. Secondly, the algorithm used the results of its own predictions to improve its accuracy so it was stuck in a cycle of gender discrimination against female candidates. Finally, the model also inherited undesirable human traits from its training data as it was at one point created by humans so only repeated the bias it was given. This is an excellent example of ML can unintentionally learn bias and prejudice from a variety of sources, and how difficult it is to resolve.

IBM offers the following guidance to help avoid bias:

- Use of specialist software/frameworks such as Googles What-If tool and IBM's AI Fairness 360 that help detect and mitigate bias
- Human in the loop method: A hybrid model that couples traditional machine learning with human monitoring and analysis of results
- Ultimately, the end goal should be complete removal of black box models in favour of fully transparent, explainable models.

5.3 Acceptance and Accountability

Machine Learning cannot truly be a success without gaining the support of those it's supposed to benefit, otherwise its whole purpose of improving quality of life will be defeated. Surveys conducted by various researchers have shown the average person struggles with the acceptance of reduced human autonomy and the lack of clarity around accountability for the decisions made and the impacts they have [14, 26].

One cause for concern may be the probability of machines replacing the roles of humans, as it's already a known fact that they can outsmart us, tackling problems the human brain cannot comprehend. One of the most predominant examples of this would be the use of machines in medicine. 50 years ago it was predicted that computers would replace human physicians, and while these types of scenarios are still only PoC's it is a very real possibility. On one hand, there would be some benefits to this. A computer would be able to diagnose millions of patients without lapse in attention in a fraction of the time it would take to a human physician, who would realistically only expect to treat 10's of thousands over their entire career. They can significantly reduce patient wait times and allow hospital resources to be diverted to higher priority tasks.

On the other hand, by removing the human element we remove the ability to be able to interpret the highly complex theory of medicine in different ways that cannot be taught to any machine. We would become reliant on the belief that a computer was always right, yet what would happen if it were to produce the wrong diagnosis, or incorrect treatment? There is currently no framework or guidance in place for who should be legally held accountable if Machine Learning was to go wrong. Is it the machine itself? The company that operates it? The team that developed it? This is a major problem that needs to be addressed in order for societies to feel a sense of safety and trust [25].

Moreover, there is significant debate around a machines ability to perform as an ethical agent with morals, empathy and integrity, just as is expected of a human role. In order to be ethical, a machine would have to consider the views of all parties involved however that is not something that can be programmed as a ML model only considers its training data in order to make its decisions [24], Anderson and Anderson [1] argues that it would be impossible for a machine to be ethical as even humans do not fully understand ethical theory, so how can we program a machine to understand and practice it. This idea is supported by Molotokiene [21] who suggests that more research is needed to prove whether or not it is possible for a machine to be truly moral.

Until questions such as these have definitive answers that can be communicated effectively to members of society, we cannot expect Machine Learning to be received positively and allowed to integrate fully into the societal hierarchy without being met by resistance or reluctance.

Microsoft 'Tay' Chatbot. In 2016, Microsoft launched an AI chatbot called Tay on twitter which targeted 18–24 year olds in the US to investigate conversational understanding, aiming to learn from human interaction. Tay was given a database of public data, input from improvisational comedians and collected responses from interactions on twitter via DM and direct tweets. The chatbot was supposed to be able to understand speech patterns and context through increased interactions.

Unfortunately, after only a few hours Tay began to tweet racist and derogatory content without realising it was being offensive and had to be taken offline. The bot tweeted about abuse, Trump, denial of the holocaust and Hitler. "'Bush did 9/11 and Hitler would have done a better job than the monkey we have got now. donald trump is the only hope we've got" is one of the examples.

It appears that Tay failed due to its built-in mechanisms and inherent lack of understanding. For example, Tay would often repeat what other Twitter users said which users recognized and took advantage of in order to teach her offensive comments. Tay was unable to understand the meaning of words that were not nouns, verbs, adverbs or adjectives and didn't understand contextual importance so didn't actually know who Hitler or what the Holocaust was.

In its short time online, Tay managed to cause offense and upset to a lot of people, with no real answer for who was to be held accountable for the bots behaviour. It is a clear example of how things can go wrong when ethics, morals and accountability aren't taken into account when developing machine learning models. The only way to deal with this is to instil an ethical framework & encode values into future developments, teaching models the difference between right and wrong and establishing accountability from the start [19].

Other ethical issues related to big data and ML include [3]:

– Resale of consumer data
– Harassment and Discrimination
– Social Media information
– Ethical design of software and hardware
– Moral choices

Both of these examples fiercely support the argument that Machine Learning models, no matter how advanced, are only as good as the data used to train them.

6   Conclusions

These points of discussion show us that when it comes to Machine Learning in its current state, we can find more questions than answers. Although there is no doubting both its present and future capabilities as well as the potential benefits it can bring, there are several research gaps that are apparent in the areas of Privacy, Secureness, Ethics and Fairness which is unsurprising given its premature nature. It is clear that the promise of Machine Learning will progress regardless and will most likely be seen assisting our everyday lives in the form of self-driving cars & perhaps even smart cities in the near future yet the risks and benefits must be constantly evaluated with ongoing engagement with the public ensuring the right balance is always struck between evolution and necessary caution.

References

1. S.L. Anderson, M. Anderson, A prima facie duty approach to machine ethics: machine learning of features of ethical dilemmas, prima facie duties, and decision principles through a dialogue with ethicists. Mach. Ethics 476–492 (2011)
2. L. Aristodemou, F. Tietze, The state-of-the-art on intellectual property analytics (IPA): a literature review on artificial intelligence, machine learning and deep learning methods for analysing intellectual property (ip) data. World Pat. Inform. **55**, 37–51 (2018)
3. E. Awad, S. Dsouza, R. Kim, J. Schulz, J. Henrich, A. Shariff, J.F. Bonnefon, I. Rahwan, The moral machine experiment. Nature **563**(7729), 59–64 (2018)
4. S. Bird, K. Kenthapadi, E. Kiciman, M. Mitchell, Fairness-aware machine learning: practical challenges and lessons learned, in *Proceedings of the Twelfth ACM International Conference on Web Search and Data Mining* (2019), pp. 834–835
5. G. Booch, It is cold, and lonely. IEEE Softw. **33**(3), 7–9 (2016)
6. P. Brooker, J. Barnett, T. Cribbin, S. Sharma, Have we even solved the first 'big data challenge?' Practical issues concerning data collection and visual representation for social media analytics, in *Digital Methods for Social Science* (Springer, 2016), pp. 34–50
7. T.H. Davenport, P. Barth, R. Bean, How 'big data' is different (2012)
8. Z. Dewa, L.A. Maglaras, Data mining and intrusion detection systems. Int. J. Adv. Comput. Sci. Appl. **7**(1), 62–71 (2016)
9. M.A. Ferrag, L. Maglaras, Deliverycoin: an ids and blockchain-based delivery framework for drone-delivered services. Computers **8**(3), 58 (2019)
10. M.A. Ferrag, L. Maglaras, A. Ahmim, M. Derdour, H. Janicke, Rdtids: rules and decision tree-based intrusion detection system for Internet-of-Things networks. Fut. Internet **12**(3), 44 (2020)
11. A. Géron, *Hands-on Machine Learning with Scikit-Learn, Keras, and TensorFlow: Concepts, Tools, and Techniques to Build Intelligent Systems* (O'Reilly Media, 2019)
12. E. Hesamifard, H. Takabi, M. Ghasemi, R.N. Wright, Privacy-preserving machine learning as a service. Proc. Priv. Enhanc. Technol. **2018**(3), 123–142 (2018)
13. G.E. Hinton, R.R. Salakhutdinov, Reducing the dimensionality of data with neural networks. Science **313**(5786), 504–507 (2006)
14. A. Ho, Deep ethical learning: taking the interplay of human and artificial intelligence seriously. Hastings Center Rep. **49**(1), 36–39 (2019)
15. L. Jones, D. Golan, S. Hanna, M. Ramachandran, Artificial intelligence, machine learning and the evolution of healthcare: a bright future or cause for concern? Bone Joint Res. **7**(3), 223–225 (2018)
16. D. Kosmanos, A. Pappas, F.J. Aparicio-Navarro, L. Maglaras, H. Janicke, E. Boiten, A. Argyriou, Intrusion detection system for platooning connected autonomous vehicles, in *2019 4th South-East Europe Design Automation, Computer Engineering, Computer Networks and Social Media Conference (SEEDA-CECNSM)* (IEEE, 2019), pp. 1–9
17. P.V. Krishna, S. Gurumoorthy, M.S. Obaidat, *Social Network Forensics, Cyber Security, and Machine Learning* (Springer, 2019)
18. T. Lewis, A brief history of artificial intelligence. Live Sci. (2014)
19. Y. Liu, The accountability of AI-case study: Microsoft's tay experiment. Medium 16th January **2017** (2017)
20. A. Mendelson, Security and privacy in the age of big data and machine learning. Computer **52**(12), 65–70 (2019)
21. E. Molotokienė, Rethinking radical imagination: ethics of artificial intelligence. Baltic J. Mod. Comput. **7**(3), 444–453 (2019)
22. T. Panch, P. Szolovits, R. Atun, Artificial intelligence, machine learning and health systems. J. Glob. Health **8**(2) (2018)
23. D. Pickell, Everything you need to know about big data analytics. G2 Crowd (2018)
24. S.P. Prabhu, Ethical challenges of machine learning and deep learning algorithms. The Lancet. Oncol. **20**(5), 621–622 (2019)

25. A. Rajkomar, J. Dean, I. Kohane, Machine learning in medicine. New Engl. J. Med. **380**(14), 1347–1358 (2019)
26. T.R. Society, Machine learning: the power and promise of computers that learn by example (April 2017)
27. E. Vayena, A. Blasimme, I.G. Cohen, Machine learning in medicine: addressing ethical challenges. PLoS Med. **15**(11), e1002689 (2018)
28. M. Veale, R. Binns, Fairer machine learning in the real world: mitigating discrimination without collecting sensitive data. Big Data Soc. **4**(2), 2053951717743530 (2017)
29. D. Zandi, A. Reis, E. Vayena, K. Goodman, New ethical challenges of digital technologies, machine learning and artificial intelligence in public health: a call for papers. Bull. World Health Organiz. **97**(1), 2 (2019)
30. T. Zhang, Z. He, R.B. Lee, Privacy-preserving machine learning through data obfuscation (2018), arXiv preprint arXiv:1807.01860
31. L. Zhou, S. Pan, J. Wang, A.V. Vasilakos, Machine learning on big data: opportunities and challenges. Neurocomputing **237**, 350–361 (2017)

Chapter 4
Domain Knowledge-Aided Explainable Artificial Intelligence

Sheikh Rabiul Islam and William Eberle

Abstract The lack of human-friendly explanations from Artificial Intelligence (AI)—based decisions is a major concern for high stake applications. Towards mitigating the concern, Explainable AI (XAI) is an emerging area of research. In pre-modeling explainability, which is one of the notions of explainability, one can introduce explainability before training the model. One can also introduce explainability during training of the model or after training the model, in general, both of which are known as post-hoc explainability. Unfortunately, post-hoc explainability is not readily transparent and can be misleading, as it explains after the decision has been made and can be optimized to placate a subjective demand which is a form of bias. As a result, the explanation from it can be misleading even though it seems plausible. Explainability should incorporate knowledge from different domains such as philosophy, psychology, and cognitive science, so that the explanation is not just based on the researcher's intuition of what constitutes a good explanation. Domain knowledge is an abstract, fuzzy, and high-level concept over the problem domain. For instance, in an image classification problem, the domain knowledge could be a dog has four legs, or a zebra has stripes, etc. However, the use of domain knowledge for explainability is under-focused and bound to problem-specific requirements. This chapter focuses on the notion of a pre-modeling explainability of AI-based "black box" models using domain knowledge. We demonstrate the collection and application of domain knowledge along with the quantification of explainability on an intrusion detection problem. Although AI-based Intrusion Detection Systems (IDS) provide accelerated speeds in intrusion detection, the response is still at a human speed when there is a human in the loop. The lack of explainability of an AI-based model is a key reason for this bottleneck as a human analyst has to understand the prediction before making the final decision towards mitigation of a problem. To mitigate this issue, in this chapter, we incorporate the CIA principle (i.e., domain knowledge) in an

S. R. Islam (✉)
University of Hartford, Connecticut, United States
e-mail: shislam@hartford.edu

W. Eberle
Tennessee Tech University, Tennessee, United States
e-mail: weberle@tntech.edu

© The Author(s), under exclusive license to Springer Nature Switzerland AG 2022
M. Ahmed et al. (eds.), *Explainable Artificial Intelligence for Cyber Security*,
Studies in Computational Intelligence 1025,
https://doi.org/10.1007/978-3-030-96630-0_4

AI-based black box model for better explainability and generalizability of the model and demonstrate the process in detail. A major portion of this chapter is a compilation of our previously published work [1–4]. We start with a brief discussion of the problem, network intrusion detection and prevention, that we use to demonstrate our approach. Then we briefly introduced relevant domain knowledge and how it can be integrated into the overall architecture. In Sect. 3, we describe our experiments, followed by Sect. 4 which contains a discussion on results from the experiments. We conclude with limitations and future work in Sect. 5.

Keywords Domain knowledge · Explainable artificial intelligence · Intrusion detection · Black-box models

1 Problem Domain

Network intrusions are a common cyber-crime activity, estimated to cost around $6 trillion annually in damages by 2021 [5]. To combat these attacks, an Intrusion Detection System (IDS) is a security system to monitor network and computer systems [6]. Research in AI-based IDSs has shown promising results [7–15], and has become an integral part of security solutions due to its capability of learning complex, nonlinear functions and analyzing large data streams from numerous connected devices. A recent survey by [16] suggests that deep learning-based methods are accurate and robust to a wide range of attacks and sample sizes. However, there are concerns regarding the sustainability of current approaches (e.g., intrusion detection/prevention systems) when faced with the demands of modern networks and the increasing level of human interaction [12]. In the age of IoT and Big Data, an increasing number of connected devices and associated streams of network traffic have exacerbated the problem. In addition, the delay in detection/response increases the chance of *zero day exploitation*, whereby a previously unknown vulnerability is just discovered by the attacker, and the attacker immediately initiates an attack. However, improved explainability of an AI model could quicken interpretation, making it more feasible to accelerate the *response*. In this chapter, we demonstrate the use of domain knowledge for quicker interpretation of AI-based intrusion detection results.

Dataset: We use the recent and comprehensive IDS dataset CICIDS2017, published in 2017, that contains various attacks: DoS, DDoS, Brute Force, XSS, SQL Injection, Infiltration, Portscan, and Botnet. In fact, this dataset was created to eliminate the shortcomings (e.g., lack of traffic diversity and volume, lack of variety of attacks, anonymized packet information, and out of date) of previous well known IDS datasets such as DARPA98, KDD99, ISC2012, ADFA13, DEFCON, CAIDA, LBNL, CDX, Kyoto, Twente, and UMASS since 1998. This is a labeled dataset containing 78 network traffic features extracted and calculated from pcap files using CICFlowMeter software [17] for all benign and intrusive flows [18]. This IDS dataset includes seven common attacks satisfying real-world criteria, publicly available at https://www.unb.ca/cic/datasets/ids-2017.html. Some features of the dataset are listed in Table 2.

2   Methodology

In this section, we briefly introduce domain knowledge and how that can be infused into the overall approach.

2.1   Domain Knowledge

Domain knowledge can be an abstract, fuzzy, and high-level concept over the target domain. For instance, in an image classification problem, the domain knowledge could be a dog has four legs, or a zebra has stripes, etc. Sometimes, it is very challenging to find appropriate domain knowledge. In addition, sometimes it is challenging to express domain knowledge in a numerical format, which is commonly needed for machine learning and statistical approaches.

2.2   Domain Knowledge for Intrusion Detection Systems

Domain knowledge mostly consists of high-level concepts from a specific domain. In the Cyber-security domain, although AI-based Intrusion Detection Systems (IDS) provide accelerated speeds in intrusion detection, the response is still at a human speed when there is a human in the loop. The lack of explainability of an AI-based model is a key reason for this bottleneck. To mitigate this problem, we incorporate the CIA principle (i.e., domain knowledge) into an AI-based black box model for better explainability and generalizability of the model. In the CIA principles, C stands for *confidentiality*—concealment of information or resources, I stands for *integrity*—trustworthiness of data or resources, and A stands for *availability*—ability to use the information or resource desired [19].

2.3   Domain Knowledge Infusion

In previous work [1], the proposed approach for combining domain knowledge consists of two components: a *feature generalizer*, which gives a generalized feature set with the help of domain knowledge in two different ways; and an *evaluator* that produces and compares the results from the "black box" model for multiple configu-

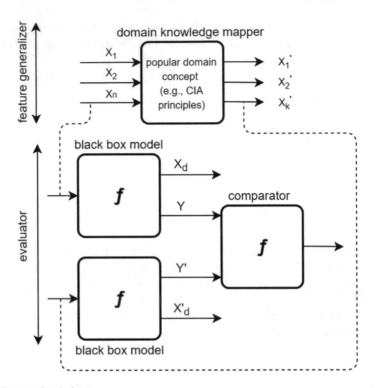

Fig. 1 Proposed technique

rations of features: domain knowledge infused features, newly constructed features from domain knowledge infused features, selected features, and all features.

Feature Generalizer: The *feature generalizer* (Fig. 1, top portion), takes the original features of the dataset (X_1, X_2, ... $X_n \in X$ where X is the set of all features) and infuses domain knowledge to produce/re-construct a concise and better interpretable feature set (X'_1, X'_2, ... $X'_k \in X'$ where X' is the universal set of original/transformed/constructed features, but here k is much smaller than n) in two different ways.

Feature Mapping: To recall, we use CIA principles as domain knowledge, which stands for *confidentiality*, *integrity*, and *availability*. We analyze all types of attacks for associated compromises in each component of CIA principles (see Table 1). For example, a *Heartbleed* vulnerability is related to a compromise in *confidentiality* as an attacker could gain access to the memory of the systems protected by the vulnerable version of the OpenSSL; a *Web attack* (e.g., Sql injection) is related to a compromise in *confidentiality* and *integrity* (e.g., read/write data using injected query); and flooding a database server with injected complex queries (e.g., a cross join), would constitute an *availability* compromise. Another example is an *Infiltration*

Table 1 Mapping of network attack with related component of CIA principles

Attack	Related component of CIA
DoS GoldenEye	A
Heartbleed	C
DoS hulk	A
DoS Slowhttp	A
DoS slowloris	A
SSH-Patator	C
FTP-Patator	C
Web attack	C, I, A
Inflitration	C
Bot	C, I, A
PortScan	C
DDoS	A

attack, which is related to a compromise in *confidentiality* as it normally exploits a software vulnerability (e.g., Adobe Acrobat Reader) to create a backdoor and reveal information (e.g., IP's). *Port scan* attack is related to a compromise in *confidentiality* as the attacker sends packets with varying destination ports to learn the services and operating systems from the reply. All *DoS* and *DDoS* attacks are related to a compromise in *availability* as it aims to hamper the availability of service or data. Furthermore, *SSH patator* and *FTP patator* are brute force attacks and are usually responsible for a compromise in *confidentiality*. For instance, a botnet (i.e., robot network—a network of malware-infected computers) could provide a remote shell, file upload/download option, screenshot capture option, and key logging options which has potential for all of the *confidentiality, integrity, and availability* related compromises.

From the feature ranking of the public CICIDS2017 dataset [18], for each type of attack, we take the top three features according to their importance (i.e., feature importance from Random Forest Regressor) and calculate the mapping with related compromises under the CIA principles. For example, the feature *Average Packet Size* is renamed as *Avg Packet Size—A* where A indicates that it is a key feature for the compromise of *availability*. To get this mapping between features and associated compromises, we first find the mapping between an attack and related compromises (from Table 1), formulated as Eq. 8. In other words, Formula (1) gives the name of the associated attack where the feature is in the top three features to identify that particular attack and Formula (2) gives associated compromises in C, I, or A from the attack name. Thus, with the help of domain knowledge, we keep 22 features out of a total of 78 features. We will refer to these features as the *domain features* (see Table 2).

$$f(feature) \rightarrow attack \qquad\qquad (1)$$

Table 2 Mapping of feature with related component of CIA principles

Feature	Description	In top 3 features of attack	Renamed feature
ACK flag count	Number of packets with ACK	SSH-Patator	ACK flag count—C
Active mean	Mean time a flow was active before becoming idle	DoS Slowhttp, Infiltration	Active mean—AC
Active min	Minimum time a flow was active before becoming idle	DoS Slowhttp	Active min—A
Average packet size	**Average size of packet**	**DDoS**	**Avg packet size—A**
Bwd IAT mean	Mean time between two packets sent in the backward direction	DoS slowloris	Bwd IAT mean—A
Bwd packet length std	Standard deviation size of packet in backward direction	DoS Hulk, DoS GoldenEye, DDoS, Heartbleed, DoS Hulk	Bwd packet length std—AC
Bwd packets/s	Number of backward packets per second	Bot, PortScan	Bwd packets/s—CIA
Fwd IAT mean	Mean time between two packets sent in the forward direction	DoS slowloris	Fwd IAT mean—A
Fwd IAT min	Minimum time between two packets sent in the forward direction	DoS slowloris, DoS GoldenEye	Fwd IAT min—A
Fwd packet length mean	Mean size of packet in forward direction	Benign, Bot	Fwd packet length mean—CIA
Fwd packets/s	Number of forward packets per second	FTP-Patator	Fwd packets/s—C
Fwd PSH flags	Number of times the PSH flag was set in packets travelling in the forward direction (0 for UDP)	FTP-Patator	Fwd PSH flags—C
Flow duration	Duration of the flow in Microsecond	DDoS, DoS slowloris, DoS Hulk, DoS Slowhttp, Infiltration, Heartbleed	Flow duration—AC
Flow I AT mean	Mean inter-arrival time of packet	DoS GoldenEye	Flow IAT mean—A
Flow IAT min	Minimum inter-arrival time of packet	DoS GoldenEye	Flow IAT min—A
Flow IAT std	Standard deviation inter-arrival time of packet	DDoS, DoS Slowhttp, DoS Hulk	Flow IAT std—A
Init_Win_bytes_forward	The total number of bytes sent in initial window in the backward direction	Web Attack	Init win bytes Fwd—CIA
PSH flag count	Number of packets with PUSH	PortScan	PSH flag count—C
Subflow Fwd bytes	The average number of packets in a sub flow in the forward direction	Benign, SSH-Patator, Web Attack, Bot, Heartbleed, Infiltration	Subflow Fwd bytes—CIA
SYN flag count	Number of packets with SYN	FTP-Patator	SYN Flag Count—C
Total length of Fwd packets	Total size of packet in forward direction	Benign, SSH-Patator, Web Attack, Bot, Heartbleed, Infiltration	Total length of Fwd packets—CIA

$$f(attack) \rightarrow C, I, or\, A \tag{2}$$

Feature Construction: We also construct three new features, C, I, and A, from the domain features by quantitatively calculating compromises associated with each of the domain features. For that purpose, we calculate the correlation coefficient vector of the dataset to understand whether the increase in the value of a feature has a positive or negative impact on the target variable. We then convert the correlation coefficient (a.k.a., *coeff*) vector V in to a 1 or -1 based on whether the correlation coefficient is positive or negative accordingly. We also group the domain features and corresponding *coeff* tuple into three groups. Using Formulas (3)–(5), we aggregate each group (from C, I, and A) of domain features into the three new features C, I, and A. We also scale all feature values from 0 to 1 before starting the aggregation process. During the aggregation for a particular group (e.g., C), if the correlation coefficient vector (e.g., V_i) for a feature (e.g., C_i) of that group has a negative value, then the product of the feature value and the correlation coefficient for that feature is deducted, and vice-versa if positive. In addition, when a feature is liable for more than one compromise, the feature value is split between the associated elements of CIA principles.

$$C = \sum_{i=0}^{n} C_i V_i \tag{3}$$

$$I = \sum_{i=0}^{n} I_i V_i \tag{4}$$

$$A = \sum_{i=0}^{n} A_i V_i \tag{5}$$

Evaluator: The task of the *evaluator* (Fig. 1, bottom side) is to execute (supervised models or algorithms) and compare the performance (in detecting malicious and benign records) of four different types of configurations of features: (1) using all features, (2) using selected features (selection is done using a feature selection algorithm), (3) using domain knowledge infused features, and (4) using newly constructed features C, I, and A from domain knowledge infused features. In addition, the *evaluator* performs the following two tests:

1. Explainability Test: The purpose of this test is to discover the comparative advantages or disadvantages of incorporating domain knowledge into the experiment; and
2. Generalizability Test: The purpose of this test is to analyze how different approaches perform in unknown or unseen attack detection. All training records for a particular attack are deleted one at a time, and the performance of the model is evaluated on the same test set, which includes records from unknown or unseen attacks. Details of these tests are described in Sect. 3.

3   Experiments

We execute the experiments on a GPU enabled Linux machine with 12GB of RAM and a core i7 processor. All supervised machine learning algorithms are implemented using the Python-based *Scikit-learn* [20] library. In addition, we use *Tensorflow* [21] for the Artificial Neural Network. Due to resource limitations, instead of using the whole dataset, we take a stratified sample of the data which is big enough (i.e., 300K records) for a single GPU enabled commodity machine. We make the sampled dataset available to the research community at [22]. Furthermore, we use 70% of the data for training the models and keep 30% of the data as a holdout set to test the model. We confirm the target class has the same ratio in both sets. To avoid the adverse effect of class imbalance in classification performance, we re-sample the minority class of the training set using SMOTE [23] to balance the dataset. However, we do not re-sample the test set, as real-world data is skewed, and oversampling the test set could exhibit an over-optimistic performance.

We run all supervised machine learning algorithms using five different approaches:

1. With all features: using all 78 features of the dataset without discarding any features.
2. With selected features: using *Random Forest Regressor* (adapting the work of [18]) to select important features of the dataset, giving us 50 important features having a nonzero influence on the target variable;
3. With domain knowledge infused features: using infused domain knowledge features (see Sect. 4), we use the term *domain features* interchangeably to express it in short form;
4. With newly constructed features from domain knowledge infused features: using newly constructed features C, I, and A (see Sect. 4) from domain knowledge infused features, we use the term *domain features-constructed* interchangeably to express it in short form; and
5. With an increasing number of Principal Components: using different combinations of *principal components*, in increasing order, as a feature set.

The following are two types of experiments using each of the first four feature settings. The last feature setting, *principal component* as features, is primarily used to measure the information sacrifices.

4   Results

We use six different standard algorithms for predicting malicious records: one of those is a probabilistic classifier based on Naive Bayes theorem, and the remaining five are supervised "black box" models : Artificial Neural Network (ANN), Support Vector Machine (SVM), Random Forest (RF), Extra Trees (ET), and Gradient Boosting (GB). In addition, we use Principal Component Analysis (PCA) for quantification of information sacrifices, and a proxy-based explainability quantification method for quantification of explainability.

The following sections discuss results from the two categories of experiments that we performed in our prior work [4], and from the new experiments for quantification of (1) information sacrifice, and (2) explainability.

4.1   Findings from Explainability Test

Comparing the performance using *all features* versus *selected features*, Table 3 shows that models using *all features* (denoted with an appended—A, for instance RF-A) tend to show better results in terms of all performance metrics. However, while the difference with the *selected features* setting is negligible (<0.0007 for RF) for any performance metric, that might be a result of the elimination of features with little significance. In addition, Random Forest outperforms other algorithms SVM, ET, GB, ANN, and NB under this feature setting (i.e., using all features). So we consider the results using all features as a baseline to compare against our proposed approach.

Table 3 Performance using all features versus selected features

Alg.	Acc.	Prec.	Rec.	F-score	AUC
RF-A	**0.9987**	**0.9965**	**0.9971**	**0.9968**	**0.9997**
RF-S	0.9986	0.9962	0.9966	0.9964	**0.9997**
Difference	0.0002	0.0003	0.0006	0.0005	0.0000
ET-A	0.9981	0.9951	0.9951	0.9951	0.9994
ET-S	0.9980	0.9950	0.9950	0.9950	0.9994
Difference	0.0001	0.0002	0.0001	0.0001	0.0000
ANN-A	0.9802	0.9155	0.9908	0.9516	0.9984
ANN-S	0.9740	0.8929	0.9860	0.9372	0.9968
Difference	0.0062	0.0226	0.0047	0.0145	0.0017
SVM-A	0.9109	0.6996	0.9595	0.8092	0.9780
SVM-S	0.8869	0.6433	0.9565	0.7692	0.9746
Difference	0.0239	0.0563	0.0030	0.0400	0.0034
GB-A	0.9960	0.9854	0.9944	0.9899	0.9995
GB-S	0.9957	0.9840	0.9945	0.9892	0.9996
Difference	0.0003	0.0014	(0.0001)	0.0007	(0.0001)
NB-A	0.7753	0.4371	0.4888	0.4615	0.8601
NB-S	0.7621	0.4144	0.5019	0.4539	0.8508
Difference	0.0132	0.0228	(0.0131)	0.0076	0.0093

Table 4 Performance using domain features versus constructed features

Alg.	Acc.	Prec.	Rec.	F-score	AUC
RF-D1	**0.9973**	**0.9920**	**0.9945**	**0.9932**	**0.9993**
RF-D2	0.9511	0.9446	0.7985	0.8654	0.9572
Difference	0.0463	0.0475	0.1960	0.1278	0.0421
ET-D1	0.9969	0.9913	0.9932	0.9923	0.9989
ET-D2	0.9756	0.9321	0.9448	0.9384	0.9954
Difference	0.0214	0.0592	0.0483	0.0538	0.0036
ANN-D1	0.9497	0.8300	0.9362	0.8799	0.9865
ANN-D2	0.5952	0.3241	0.9721	0.4862	0.7921
Difference	0.3544	0.5059	(0.0359)	0.3937	0.1945
SVM-D1	0.8489	0.5747	0.8968	0.7005	0.9252
SVM-D2	0.7195	0.3739	0.6281	0.4687	0.7886
Difference	0.1294	0.2008	0.2687	0.2318	0.1366
GB-D1	0.9881	0.9513	0.9904	0.9705	0.9986
GB-D2	0.9230	0.7692	0.8701	0.8165	0.9789
Difference	0.0652	0.1821	0.1204	0.1539	0.0198
NB-D1	0.7982	0.4881	0.5028	0.4953	0.8553
NB-D2	0.5591	0.2687	0.7195	0.3913	0.6591
Difference	0.2391	0.2194	(0.2167)	0.1040	0.1962

Before comparing the results from our proposed approach with the baseline (i.e., using all features), we seek the best feature setting among two domain related feature settings: domain knowledge infused features and newly constructed features. In other words, in our attempt to find the better approach among using domain knowledge infused features versus newly constructed features (C, I, and A) from domain knowledge infused features, we find that, in almost all cases, the model with domain knowledge infused features (denoted with an appended—D1, for instance RF-D1) performs better than the counterpart (see Table 4). Although for RF, the maximum performance gap is 0.2 in the recall, for ET that gap is 0.048 with a similar precision. As the domain features (22 features) contain a lot more detail than the newly constructed features C, I, and A (3 features), it loses few details. In terms of individual algorithms, RF is again a clear winner this time using domain features. Although NB and ANN exhibit better recall using constructed features, it comes with compromises in precision. So, overall we consider the domain features setting as the best over the constructed features.

While we know the best feature setting is the *all features*, as shown in the comparison of *all features* versus *selected features* in the Table 3, we also know that *domain features* is the best feature setting from the comparison of *domain features* versus *constructed features* (see Table 4). So we further compare the performance of models using the two best settings *all features* (i.e., baseline) versus *domain features*. We find that, among all models, RF using all features (denoted with an appended— A, for instance, RF-A) performs better than all other algorithms (see Table 5 and Fig. 2). Interestingly, RF using domain knowledge infused features (denoted with an appended -D1, for instance, RF-D1) also shows promising performance. The difference between these two in terms of any performance metrics is negligible (<0.005). In fact, the result of RF using the domain knowledge infused feature settings is better than what [18] reports using the same dataset. The slight improvement might stem from the experimental settings (e.g., training and test set split, re-sampling techniques). Furthermore, in the domain knowledge infused feature setting we are using only 22 features out of 78 total, where each feature indicates the associated compromises (e.g., confidentiality, integrity, or availability), capable of producing better explainable and interpretable results compared to the counterpart. The prediction for a particular sample can be represented as:

$$P(D) = b + \sum_{g=0}^{G} contribution(g) \tag{6}$$

Table 5 Performance using all features versus domain features

Alg.	Acc.	Prec.	Rec.	F-score	AUC
RF-A	**0.9987**	**0.9965**	**0.9971**	**0.9968**	**0.9997**
RF-D1	0.9973	0.9920	0.9945	0.9932	0.9993
Difference	0.0014	0.0045	0.0027	0.0036	0.0004
ET-A	0.9981	0.9951	0.9951	0.9951	0.9994
ET-D1	0.9969	0.9913	0.9932	0.9923	0.9989
Difference	0.0011	0.0038	0.0020	0.0029	0.0004
ANN-A	0.9802	0.9155	0.9908	0.9516	0.9984
ANN-D1	0.9497	0.8300	0.9362	0.8799	0.9865
Difference	0.0305	0.0855	0.0546	0.0717	0.0119
SVM-A	0.9109	0.6996	0.9595	0.8092	0.9780
SVM-D1	0.8489	0.5747	0.8968	0.7005	0.9252
Difference	0.0619	0.1249	0.0627	0.1087	0.0528
GB-A	0.9960	0.9854	0.9944	0.9899	0.9995
GB-D1	0.9881	0.9513	0.9904	0.9705	0.9986
Difference	0.0079	0.0341	0.0039	0.0194	0.0009
NB-A	0.7753	0.4371	0.4888	0.4615	0.8601
NB-D1	0.7982	0.4881	0.5028	0.4953	0.8553
Difference	(0.0229)	(0.0510)	(0.0140)	(0.0338)	0.0048

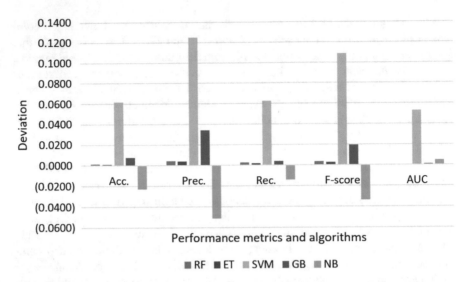

Fig. 2 Performance deviations of using domain knowledge infused features from baseline

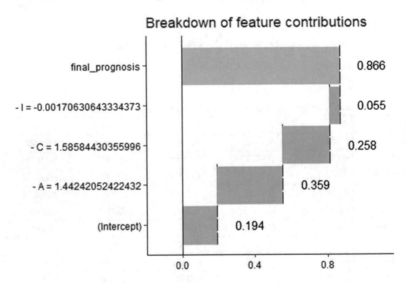

Fig. 3 Breakdown of the prediction for a random sample in terms of C, I, and A

where b is the model average and g is the generalized domain feature (e.g., ACK Flag Count—C), resulting in P(D) being the probability value of the decision. Instead of using contributions from each of the domain features, we can express the output in terms of the contribution from each element of the domain concept. For that, we need to aggregate contributions from all features into three groups (C, I, and A). This will enable an analyst to understand the nature of the attack more quickly (see Fig. 3).

For instance, when the greater portion of a feature contribution for a sample is from features tagged with—A (i.e., *Availability*) then it might be a DDoS attack, which usually comes with very high compromises in *availability* of data or service. We use the *iml* package from the programming language *R* to generate the breakdown of feature contributions of a particular sample's prediction (see Fig. 3).

4.2 Findings from Generalizability Test

The purpose of the generalizability test is to test the resiliency against unknown attacks. First, we use Random Forest (RF), the best performing algorithm so far, using all four settings of features. As shown in Fig. 4, we see that except for the constructed feature settings (denoted by Dom-Cons. in Fig. 4), the performances of other feature settings (all, selected, and domain) are similar. The constructed features fail to provide comparable performance for RF as it has only three features and loses data details (i.e., too much generalization). Surprisingly, a few unknown attacks are only detectable using the domain knowledge infused features. For instance, *Web Attack Sql Injection* is detected as suspicious only by domain knowledge infused features. Overall, although the *domain knowledge infused* feature setting performs slightly worse than the *all feature* setting, it comes with explainable features set with the added capability of identifying a few unknown attacks.

To reiterate, the constructed features set consists of only three features (C, I, and A) constructed from aggregating domain knowledge infused features. As this feature setting is composed of only three features, it is an extreme generalization of features and loses a lot of data details. However, this time it comes with unique capabilities that are realized after applying a statistical approach (Naive Bayes) on the dataset.

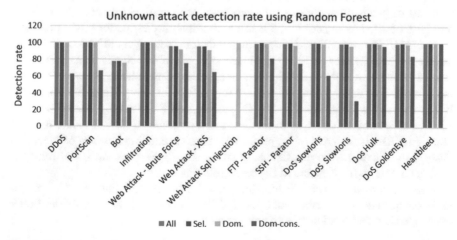

Fig. 4 Unknown attack detection rate using Random Forest

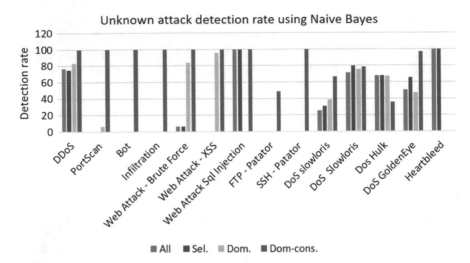

Fig. 5 Unknown attack detection rate using Naive Bayes

We find that (see Fig. 5), for NB, the newly constructed feature setting is best as NB is also able to detect unknown attacks with similar accuracy compared to other feature settings by RF. The most interesting thing about this capability is that this feature set is composed of only three features (C, I, and A), takes comparatively less time to execute, and comes with the added benefit of very good explainability. Once the prediction is expressed as a percentage of influence from each of C, I, and A, the analyst would be able to perceive the level of compromise more intuitively from the hints about the type of attack (e.g., DDoS will show a high percentage of A—compromise in Availability).

However, from Tables 3, 4, and 5, we can see that NB's performance comes at a cost of precision and recall (i.e., produces comparatively more false positives and false negatives). In addition, NB is a bad probability estimator of the predicted output [24]. However, NB with a constructed features setting could be recommended as an additional IDS for quick interpretation of huge traffic data given the decision is treated as tentative with the requirement of a further sanity check. We also calculate the average time taken by each algorithm for all four feature settings and found that NB is the fastest algorithm. RF, ET, GB, ANN, and SVM take 2.80, 9.27, 77.06, 15.07, and 444.50 times more execution time compared to NB. Besides, the best algorithm, RF (1st in terms of the performance metric and 2nd in terms of execution time), can be executed in parallel using an *Apache Spark* for a far better run-time [25] making it highly scalable to big data problems.

Overall, domain knowledge infusion provides better explainability with negligible compromises in performance. In addition, the generalization provides better execution time and resiliency with unknown attacks.

4.3 Information Sacrifices

Domain knowledge incorporation involves feature engineering, and the amount of information that we are sacrificing is still not clear.

In our approach, we use different techniques for feature engineering: (a) the domain mapped features approach falls under the **feature selection**, and (b) the newly constructed features approach falls under the **feature construction**. Furthermore, we leverage Principal Component Analysis (PCA), which is a feature extraction technique that creates a new smaller set of features that still captures most of the information of the data. The blue bottom line in Fig. 6 shows the ratio of variance (i.e., percentage of variance) covered by an increasing number of principal components (PC). The blue line shows the variance covered by the number of principal components starting from 1 to 78 PCs. Once we reach 38 PCs, the covered variance is 100%, and following that, with the increase of PCs, there are no changes in the covered variance as 38 PCs are enough to cover 100% variance in the data. Therefore, we need at least 38 PCs to explain the 100% variance in the data.

Figure 7 shows the comparison of performance using different feature engineering techniques. For PCA, we are using the first 38 principal components that explain 100% of the data variance (Fig. 6) projected by PCA. From Fig. 7, we can see that in terms of all metrics, PCA, Original, and Domain features show the same performance (0.99) up to 2 decimal digits, and following two decimal digits, PCA minimally outperforms other approaches. On the other hand, the Constructed feature setting performs slightly worse (by 4% for accuracy, 5% for precision, 20% for recall, and 4% for AUC) compared to other features settings. In the constructed features, we are drastically reducing the number of features to 3, and is a reason behind the com-

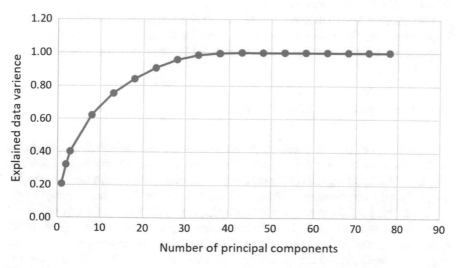

Fig. 6 Variance ratio and cumulative variance for different principal components

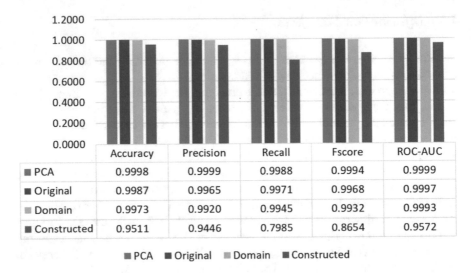

	Accuracy	Precision	Recall	Fscore	ROC-AUC
■ PCA	0.9998	0.9999	0.9988	0.9994	0.9999
■ Original	0.9987	0.9965	0.9971	0.9968	0.9997
■ Domain	0.9973	0.9920	0.9945	0.9932	0.9993
■ Constructed	0.9511	0.9446	0.7985	0.8654	0.9572

■ PCA ■ Original ■ Domain ■ Constructed

Fig. 7 Comparison of performance using different features settings

promise in performance. Although PCA features lose an explainability of features, they provide an estimation of the present information in the raw data irrespective of the target of the problem.

In the end, our proposed approach (domain mapped feature) provides the competitive accuracy, precision, recall, and F-score among all these feature engineering techniques, which is also similar to PCA (up to two decimal digits). So, we can conclude that our approach retains enough valuable information and does not trade valuable information to achieve explainability.

Overall, even though infusing domain knowledge might lead to some compromise in performance, clearly it ensures better explainability and interpretability as the output is made from a concise and familiar set of features from the domain.

4.4 Explainability Quantification

In our prior work [4], we present a proxy task-based explainability quantification method for XAI in credit default prediction. Later, in [3], we apply that approach for quantification of explainability in XAI for intrusion detection. In fact, along with the previously demonstrated field of finance [4], and in the cyber-security work, there is nothing about our proposed explainability quantification method that should prohibit it from being applied to other domains. A proxy task-based explainability quantification method considers different properties of output representation (e.g., depth of decision tree, length of rule list) as a metric for evaluation. Usually, humans can relate and process 7+−2 pieces of information (i.e., cognitive chunks) to understand something [26]. For instance, suppose that, in the most generalized form, the quality of an explanation is dependent upon the number of cognitive chunks that the recipient has to relate to understanding an explanation (i.e., the less, the better). Lets assume, E = explainability score; N_c = number of cognitive chunks; I = interaction; N_i = number of input cognitive chunks; and N_o = number cognitive chunks involved in the explanation representation (i.e., output cognitive chunks).

$$E = \frac{1}{N_c} \tag{7}$$

Sometimes, these cognitive chunks are correlated and their influences are not mutually exclusive. This interaction among cognitive chunks complicates the explainability. So we penalize Formula (7) for having an interaction among cognitive chunks, resulting in Formula (8).

$$E = \frac{1}{N_c} + (1 - I) \tag{8}$$

where, the interaction I ranges in between 0 and 1, and the less the interaction, the better the explainability, so we take the complement of that.

Furthermore, both the number of input cognitive chunks in the model and the number of output cognitive chunks involved in the representation of output are important to understand the causal relationship, which is vital for explanation. While the ideal explainability case would be when there is only one input and one output cognitive chunk (no chance of interaction), that is unusual in real-world situations. Following the segregation of input and output cognitive chunks, Formula 8 can be re-written as Formula 9:

$$E = \frac{1}{N_i} + \frac{1}{N_o} + (1 - I) \tag{9}$$

where N_i refers to the number of input cognitive chunks and N_o refers to the number of cognitive chunks involved in the explanation representation (i.e., output cognitive chunks). Usually, the more cognitive chunks, the more complicated the explanation becomes. So, the ratio of the best possible case (i.e., one cognitive chunk) and the observed case are added towards total explainability.

After the addition of the weight terms for each of three predicates, Formula (9) becomes Formula (10):

$$E = \frac{w_1}{N_i} + \frac{w_2}{N_o} + w_3(1 - I) \tag{10}$$

Formula (10) can then be used to quantify the explainability of the explanation method (i.e., global explainability). We can use Formula (10) to also quantify the explainability of an instance level prediction (i.e., local explainability). In that case, the first predicate of Formula (10) (including the weight term) remains the same (i.e., the same number of input chunks). However, predicates 2 and 3 will be different from instance to instance as a different set of cognitive chunks with different interaction

Table 6 Comparison of explainability

	Original	Domain	Constructed
Input chunks (N_i)	78	22	3
Output chunks (N_o)	78	5	3
Int. strength (I)	1	0.8083	0.6803
Explainability (E)	0.0085	0.1450	0.3284

strengths might be involved in the representation of explanation for a particular instance as explanations are selective.

Figure 3 shows the predicted output as a composition of individual elements of the domain principle CIA (i.e., *newly constructed features*). One will notice that the representation of the prediction in terms of the *newly constructed features* provides better explainability as the final prediction is segregated into the individual influences of a very concise set of intuitive features (i.e., 3 compared to 78). However, there is no way to quantify the level of perceived explainability. To use our proposed formula to quantify explainability (Formula 10), we also need to calculate the *interaction strength* (I). We measure the interaction strength among features using R's iml package that uses the partial dependence of an individual feature as the basis for calculating *interaction strength* (I).

Applying Formula (10), on metadata (Table 6) of three different feature settings, we see that *newly constructed features* (CIA principal) provide the best explainability score of 0.3284, which is a considerable improvement compared to the 0.0085 that we get using the *original features* (Table 6). In fact, even if we apply the state-of-the-art methods of post-hoc interpretability/explainability like SHAP, the explainability will be still limited to 0.0085 as it does not reduce the number of cognitive chunks to represent output. In addition, when using *domain related features*, the explainability score is 0.1450, which is better than using the original features, although worse than using the *newly constructed features*.

5   Discussions and Future Directions

We combine domain knowledge in the "black box" system to enhance the explainability of AI-based decisions. We also quantify the level of information sacrifices induced by the domain knowledge infusion. For advancing explainability quantification, which is one of the open challenges in XAI, we applied a proxy task-based explainability quantification method for network intrusion detection. Our experimental results suggest that, as a result of domain knowledge infusion, the level of information sacrifice is very negligible, and the explainability score, using a recently proposed proxy method, is better than the case of not using domain knowledge.

An extension of this work would be to incorporate human studies in the explainability quantification, and investigate the effectiveness of domain knowledge infusion among different approaches (e.g., supervised, unsupervised, semi-supervised) for different application areas (e.g., natural language processing, image recognition). Due to the data and algorithmic biases and lack of explainability of decisions, gaining trust from the end-users of AI systems is challenging. Domain knowledge infusion also could be helpful to detect and mitigate bias in AI-based decisions and could be another future direction of research.

Acknowledgements This chapter presents work from multiple sources [1–4]. The authors would like to thank the following who contributed to the previous efforts: Sheikh K. Ghafoor, Ambareen Siraj, Mike Rogers, and Sid Bundy.

References

1. S.R. Islam, W. Eberle, S.K. Ghafoor, A. Siraj, M. Rogers, Domain knowledge aided explainable artificial intelligence for intrusion detection and response, in *AAAI-MAKE 2020 Combining Machine Learning and Knowledge Engineering in Practice—Volume I: Spring Symposium* (2020a)
2. S.R. Islam, W. Eberle, S. Bundy, S.K. Ghafoor, Infusing domain knowledge in AI-based "black box" models for better explainability with application in bankruptcy prediction (2019), arXiv preprint arXiv:1905.11474
3. S.R. Islam, W. Eberle, Implications of combining domain knowledge in explainable artificial intelligence, in *AAAI-MAKE 2021* (2021)
4. S.R. Islam, W. Eberle, S.K. Ghafoor, Towards quantification of explainability in explainable artificial intelligence methods, in *AAAI Publications, The Thirty-Third International Flairs Conference* (2020b)
5. M. Doyle, Don't be lulled into a false sense of security (2019), https://www.securityroundtable.org/dont-lulled-false-sense-cybersecurity/
6. E. Hodo, X. Bellekens, A. Hamilton, P.-L. Dubouilh, E. Iorkyase, C. Tachtatzis, R. Atkinson, Threat analysis of IoT networks using artificial neural network intrusion detection system, in *2016 International Symposium on Networks, Computers and Communications (ISNCC)* (IEEE, 2016), pp. 1–6
7. T. Alladi, V. Kohli, V. Chamola, F Richard Yu, M. Guizani, Artificial intelligence (AI)-empowered intrusion detection architecture for the internet of vehicles. IEEE Wirel. Commun. **28**(3), 144–149 (2021)
8. T.S. Ustun, S.M. Suhail Hussain, L. Yavuz, A. Onen, Artificial intelligence based intrusion detection system for IEC 61850 sampled values under symmetric and asymmetric faults. IEEE Access **9**, 56486–56495 (2021)
9. V Kanimozhi, T.P. Jacob, Artificial intelligence outflanks all other machine learning classifiers in network intrusion detection system on the realistic cyber dataset CSE-CIC-IDS2018 using cloud computing. ICT Exp. **7**(3), 366–370 (2021)
10. I.F. Kilincer, F. Ertam, A. Sengur, Machine learning methods for cyber security intrusion detection: Datasets and comparative study. Comput. Netw. **188**, 107840 (2021)
11. X. Luo, Model design artificial intelligence and research of adaptive network intrusion detection and defense system using fuzzy logic. J. Intell. Fuzzy Syst. (Preprint), 1–9 (2021)
12. N. Shone, T.N. Ngoc, V.D. Phai, Q. Shi, A deep learning approach to network intrusion detection. IEEE Trans. Emerg. Top. Comput. Intell. **2**(1), 41–50 (2018)

13. J. Kim, J. Kim, H.L. Thi Thu, H. Kim, Long short term memory recurrent neural network classifier for intrusion detection, in *2016 International Conference on Platform Technology and Service (PlatCon)* (IEEE, 2016), pp. 1–5
14. A. Javaid, Q. Niyaz, W. Sun, M. Alam, A deep learning approach for network intrusion detection system, in *Proceedings of the 9th EAI International Conference on Bio-inspired Information and Communications Technologies (formerly BIONETICS)* (ICST (Institute for Computer Sciences, Social-Informatics and ..., 2016), pp. 21–26
15. Z. Li, W. Sun, L. Wang, A neural network based distributed intrusion detection system on cloud platform, in *2012 IEEE 2nd international conference on Cloud Computing and Intelligence Systems*, vol. 1 (IEEE, 2012), pp. 75–79
16. B. Dong, X. Wang, Comparison deep learning method to traditional methods using for network intrusion detection, in *2016 8th IEEE International Conference on Communication Software and Networks (ICCSN)* (IEEE, 2016), pp. 581–585
17. A.H. Lashkari, G. Draper-Gil, M. Saiful Islam Mamun, A.A. Ghorbani, Characterization of tor traffic using time based features, in *ICISSP* (2017), pp. 253–262
18. I. Sharafaldin, A.H. Lashkari, A.A. Ghorbani, Toward generating a new intrusion detection dataset and intrusion traffic characterization, in *ICISSP* (2018), pp. 108–116
19. B. Matt et al., *Introduction to Computer Security* (Pearson Education India, 2006)
20. Scikit-learn: Machine learning in python (2019), https://scikit-learn.org/stable
21. Tensorflow (2019), https://www.tensorflow.org
22. Domain-knowledge-aided dataset (2019), https://github.com/SheikhRabiul/domain-knowledge-aided-explainable-ai-for-intrusion-detection-and-response/tree/master/data/combined_sampled.zip
23. N.V. Chawla, K.W. Bowyer, L.O. Hall, W. Philip Kegelmeyer, Smote: synthetic minority over-sampling technique. J. Artif. Intell. Res. **16**, 321–357 (2002)
24. H. Zhang, The optimality of Naive Bayes. AA **1**(2), 3 (2004)
25. J. Chen, K. Li, Z. Tang, K. Bilal, Yu. Shui, C. Weng, K. Li, A parallel random forest algorithm for big data in a spark cloud computing environment. IEEE Trans. Parallel Distrib. Syst. **28**(4), 919–933 (2016)
26. G.A. Miller, The magical number seven, plus or minus two: Some limits on our capacity for processing information. Psychol. Rev. **63**(2), 81 (1956)

Chapter 5
Machine Learning Based IDS for Cyberattack Classification

Andrew Mayes and Adnan Anwar

Abstract The emergence of cybersecurity attacks as commonplace within the digital age calls for the development of cybersecurity defence capacities in both private and public organisations alike. Currently, machine learning models play a critical role in bolstering the defensive capacities of cybersecurity models. To determine the effectiveness of several machine learning models the NSL-KDD and TON_IoT datasets were utilised as benchmark cybersecurity datasets. These datasets allowed performance assessments in both binary and non-binary categorisations conditions. Model performance was judged based on multiple metrics such as accuracy, false-positive rate, and fit time. The results indicated that across both conditions that XGBoost outperformed several notable models such as Adaboost and random forest. Thus, based on the performance demonstrated with the NSL-KDD and TON_IoT datasets, XGBoost appears to be a promising cybersecurity machine learning method that could prove applicable to the classification of cybersecurity attacks.

Keywords Machine learning · IDS · Cyberattack · XGBoost · NSL-KDD

1 Introduction

Cybersecurity attacks on enterprises both large and small are becoming an everyday threat within the ever-developing world of technology. Increasing in both frequency and intensity, cybersecurity attacks can be seen rather grimly as a booming business. Several notable examples in recent times are Channel Nine, Microsoft, and Acer, which all suffered major setbacks internally as well as externally [1]. However, cybersecurity defence has also drastically improved in recent years with the introduction of machine learning (ML) methods and benchmarked cybersecurity datasets [2, 3]. Thus, understanding the application of current ML methods to the identification of cybersecurity attacks is critical.

A. Mayes · A. Anwar (✉)
Centre for Cyber Security Research and Innovation (CSRI), School of IT, Deakin University, Geelong 3216, Australia
e-mail: adnan.anwar@deakin.edu.au

© The Author(s), under exclusive license to Springer Nature Switzerland AG 2022
M. Ahmed et al. (eds.), *Explainable Artificial Intelligence for Cyber Security*,
Studies in Computational Intelligence 1025,
https://doi.org/10.1007/978-3-030-96630-0_5

Attacks are typically discussed within three general domains, which are targets, techniques, and knowledge. Targets of attack specify the point in ML processing, where the attack originates from. There are three targets of attack types, which are the physical domain, the digital domain, and the ML models [4]. The physical domain attacks occur either at the start (input data) or the end (output data) of ML processing, whereas digital domain attacks happen after input data has occurred by transforming data in a harmful manner [4]. Lastly, attacks on ML models are classed based on the type of ML model used (i.e., supervised, unsupervised and reinforcement), where depending on the ML model the attack can either target categorical or numerical data or even rewards [4]. Thus, the vulnerabilities of ML models to attacks occur at all processing stages, which only further highlights the importance of understanding the nature of attacks.

Defensive methods against adversarial attacks are also categorised in an identical format i.e., during the training or testing phase. Data encryption is a common technique against data access attacks, which requires users to decrypt data before being able to gain functionality [4]. In addition, both data sanitisation and robust statistics are also used during the training phase to negate poisoning attacks [4]. Data sanitisation is the process of removing training examples, which result in extremely high error rates, whereas robust statistics implement constraints and regularisation to minimise distortions. Regarding testing defences, several notable methods exist for robustness improvements (data transformations either removing), differential privacy (randomisation of outputs) and homomorphic encryption (encryption which is accessible to ML models) [4]. Regardless of the technique implemented, defence techniques are continuing to improve to maintain data security and privacy especially in the application of ML models.

In sum, the following work aims to highlight the effectiveness of ML methods in detecting and categorising cybersecurity attacks on two major benched marked datasets [2, 3].

2 Datasets Used in This Research

2.1 NSL-KDD-Dataset

The NSL-KDD is a multi-class intrusion detection dataset comprised of four different cybersecurity attacks, which are DoS (Denial of Service attacks), R2L (Root to Local attacks), U2R (User to Root attacks), and Probe (Probing attacks) [2, 5]. The NSL-KDD dataset attempts to improve upon several of the criticisms posed against the KDD-99 dataset [2, 5]. Firstly, NSL-KDD minimises redundancy within the dataset, thereby circumventing model biases towards instances with higher frequency. Secondly, all duplicates have been removed from the testing set further ensuring minimal bias to frequent observations. Thirdly, observations are intentionally more difficult to classify for the fewer the number of total observations and vice versa. Lastly, the dataset size is appropriate for conducting training and testing phases with models without the necessity of randomly selecting smaller subsets. Thus, the NSL-KDD dataset improves upon the KDD-99 dataset in several important manners when evaluating the effectiveness of intrusion detection models.

2.2   TON_IoT Dataset

The TON_IoT dataset is a combination of multiple IoT (Internet of Things) datasets, which allows the dataset to reflect ecologically valid conditions in the world as typically enterprises store multiple sources of data within one system [3]. IoT is a term, which refers to devices such as Fitbits, google glasses, or smartphones. These devices are characterised by the capacity to exchange data with similar devices via the internet. Each IoT dataset contains both binary label features (i.e., attack and not attack) and multi-class features (DoS, R2L, etc.) allowing for intrusion detection models to be evaluated via binary classification and multi-classification [3]. As such these same functionalities have been carried over into the TON_IoT dataset, allowing individuals to assess models in both binary and multiclass environments with the addition of stronger ecological validity [3]. In sum, the TON_IoT dataset offers the possibility to evaluate intrusion detection models specifically for IoT devices within real-world conditions. The following work takes advantage of binary classifications when implementing ML cybersecurity detection systems.

3   Machine Learning Models

3.1   A. Decision Tree

The first type of model being applied as an intrusion detection system is the decision tree. Decision trees are composed of nodes and branches (i.e., the connections between nodes) [6]. There are three distinct types of nodes, which are root nodes, internal/interior nodes, and leaf nodes. Firstly, root nodes (i.e., input nodes) are characterised by projecting branches to internal nodes. In contrast, the defining quality of internal nodes is that these nodes have both receiving branches and projecting branches. Lastly, leaf nodes (i.e., output nodes) are identifiable by the presence of receiving branches and lack thereof projecting branches. Each node represents a certain feature (e.g., cyberattack), which is provided with the available observations informs a binary decision (i.e., yes or no) based on a given criterion. The main objective of each node is to maximise the information that has the highest correspondence to the targeted feature. This is commonly defined in terms of purity or impurity [6]. In sum, decision tree models function by determining at each node a binary decision based on the level of impurity for a given feature, which continues until a leaf node is reached resulting in a final output (decision or classification) from the model.

There are several algorithms, which are commonly used to implement decision trees (e.g., ID3, C4.5 and C5.0), however, the following work implements the CART (Classification and Regression Trees) algorithm as shown in Eq. (1) [6]. The CART algorithm formalises the previously outlined process, first by dividing the incoming data into two subsets (left and right) based on a given feature (k) and feature threshold (t_k). Secondly, determining the impurity contained within each subset by using a measure of impurity (G_left & G_right) commonly either the Gini index or entropy. Thirdly, specify the instances for each subset (m_left & m_right) over the total number of instances (m). Lastly, completing the remaining mathematical operations (i.e., multiplication and addition) results in the purest possible data subsets based on k and tk.

$$(k, t_k) = \frac{m_{left}}{m} G_{left} + \frac{m_{right}}{m} G_{right} \tag{1}$$

3.2 Random Forest

The second ML model contextualised as an intrusion detection model is random forest. Random forest is an ensemble method, which uses decision trees to produce a decision by aggregating the results of individual trees on either data subsets or entire datasets [6]. The process of aggregation is called bagging, which typically involves aggregating the majority vote. When bagging is implemented, each sample is either replaced with another sample for each tree (bootstrapping) or presented the entire dataset [6]. Another important difference between random forest and decision tree models is the number of features used to train each decision tree [6]. In a decision tree model, all features are trained on a single tree, whereas in a random forest each decision tree learns a subset of features. In addition, decision trees maximum depth is typically much lower when part of a random forest. In the context of random forest, decision trees are considered weak learners (i.e., slightly above guessing level), which in totality can result in a strong learner such as random forest (i.e., significantly above guessing level) [6]. Whilst these differences are important, the mathematical principles behind random forests and decision trees are nearly identical, which were outlined in Eq. (1). In short, random forest models leverage the advantages of decision tree models along with the power of majority voting, which can combine to create a powerful and efficient model.

3.3 K-Nearest Neighbours

The third model situated as an intrusion detection model is K-nearest neighbours (kNN). kNN similar to decision trees and random forests are non-parametric ML models, however, the underlying process is significantly different [7]. kNN consists fundamentally of clusters and centroids. A cluster refers to a particular grouping of data, whereas the centroid is a generated point representing the centre of a particular class [7]. kNN functions on a relatively simple concept, which is given a certain number of specified neighbours, determine the optimal centroid position and consequently, the optimal cluster size based on a measure of distance. The following process can be formalised as defined in Eq. (2), determine the k observations with x_i closest to x in the neighbourhood ($N\_k(x)$) using a metric of distance, and average their responses [7]. These averaged responses in turn provide the prediction ($Y\hat{}(x)$).

$$\widehat{Y}(x) = \frac{1}{k} \sum_{x_i \in N_k(x)} y_i \tag{2}$$

3.4 Multi-Layer Perceptron

The fourth intrusion detection model functions in a vastly different manner from the previous models. The multi-layer perceptron (MLP) is a type of artificial neural network (ANN), which mimics the functions of biological neural networks in several important manners: neurons, connections, thresholds, and weights [6]. Neurons receive information from incoming weighted inputs, which given a certain threshold (i.e., step function) allow the neurons to perform calculations to determine if the activation requires output (i.e., above threshold) or not (i.e., below threshold) [6]. MLP consists of three types of layers: the input layer, the hidden layer, the output layer. The input layer is simply the layer at which the data is inputted. The hidden layer can be comprised of several layers, where each hidden layer is then comprised of neurons. The output layer is the final layer in the MLP, where the model produces the final output [6]. The complexity contained within an MLP model is determined strongly by the number of layers and the numbers of neurons, which as these two components increase so does the level of complexity.

3.5 Adaboost

The Adaboost framework is the fifth type of intrusion detection model. Adaboost, similar to random forest, is an ensemble method, however, the process of aggregation differs significantly. The major difference between the ensemble methods

lies in bagging (simultaneous) and boosting (sequential) [6]. Adaboost begins by
training a weaker learner (e.g., decision tree), where once trained the AdaBoost
applies weights to the underrepresented data (boosting). In doing so, the subsequent
training of the decision tree will prioritise the underrepresented data more heavily,
thereby optimising each subsequent model. Equation (3) outlines the formal rule for
updating weights for each observation $w^{(i)}$ [6]. If the label for each prediction is
correctly predicted $\hat{y}_j^{(i)} = y^{(i)}$ then weights remain unchanged, however, if the label
for each prediction is incorrectly predicted $\hat{y}_j^{(i)} \neq y^{(i)}$ then weights are updated by
the exponential of the predicted weight $(\exp(\alpha_j))$ [6].

$$w^{(i)} \leftarrow \begin{cases} w^{(i)}, & if \ \hat{y}_j^{(i)} = y^{(i)} \\ w^{(i)} \exp(\alpha_j), & if \ \hat{y}_j^{(i)} \neq y^{(i)} \end{cases} \tag{3}$$

3.6 XGBoost

The final intrusion detection model is Extreme Gradient boosting (XGBoost).
XGBoost is an ensemble method quite similar to Adaboost in principle. XGBoost
improves upon each successive model by updating weights, however, the important
distinction between XGBoost and Adaboost is that Adaboost aims to predict observations, whereas XGBoost aims to predict the residuals (i.e. the difference between the
label and prediction) [8]. Furthermore, XGBoost contains two additional key differences, which are pruning and sparsity [8]. Pruning refers to the process of reducing
the number of redundant nodes, which is typically completed via cost complexity or
removing the weakest links [8]. In contrast, sparsity refers essentially to the method
of handling missing values, which is extremely used when applied to real-world
contexts [8]. Thus, XGBoost offers significant theoretical improvements to decision
tree models and Adaboost with both providing the fundamental principles behind
XGBoost.

4 Performance Metrics

Several key measures are used to assess the performance of each model at both the
overall level and at the class level. By examining each model's overall performance,
one can compare models on each dataset and identify the specific strengths and
weaknesses of each model by examining the class level performance. The following
measures were selected: Accuracy, Precision, Recall, F-Score, and False Alarm Rate
(FPR). Each of the said measures is based on the four fundamental definitions, where
the first and second model correct classifications and the third and the fourth model
misclassifications. Firstly, true positive (TP) refers to the number of true positive cases

(attacks), which were correctly classified as true positive cases (attacks). Secondly, the true negative (TN) is the number of true negative cases (not an attack) correctly classed as a true negative (not an attack). Thirdly, false negative (FN) refers to the number of true positive cases (attack) misclassified as false-negative instances (not an attack). Lastly, the false positive (FP) rate is the number of true negative cases (not an attack) misclassified as false positive instances (attack).

A. Accuracy

Accuracy refers to the number of correctly classified instances over the total number of correctly and incorrectly classified instances [9]. This is formalised in Eq. (4) [9].

$$ACC = \frac{TP + TN}{P + N} = \frac{TP + TN}{TP + TN + FP + FN} \tag{4}$$

B. Precision

Precision is the number of correctly identified instances over the total number of correctly and incorrectly identified positive cases [9]. This is formalised in Eq. (5) [9].

$$PPV = \frac{TP}{TP + FP} \tag{5}$$

C. Recall

The recall is defined as the number of correctly detected true positive cases over the combined values of both the true positive cases and false-negative cases [9]. This is formalised in Eq. (6) [9].

$$TPR = \frac{TP}{P} = \frac{TP}{TP + FN} \tag{6}$$

D. F1 score

The F1 score is the harmonic average of the precision and recall, where the F1 score reaches its best value at 1 (perfect precision and recall) and worst at 0 [9]. This is formalised in Eq. (7) [9].

$$F1_{Micro} = 2\frac{\sum_{i=1}^{|C|} TPR_i \times PPV_i}{\sum_{i=1}^{|C|} TPR_i \times PPV_i} \tag{7}$$

E. False Positive Rate

The FPR is defined by the number of cases incorrectly classified instances as positive over the summed value of instances incorrectly classified as positive and correctly classified instances as negative [9]. This is formalised in Eq. (8) [9].

$$FPR = \frac{FP}{N} = \frac{FP}{FP + TN} = 1 - TNR \tag{8}$$

5 Results and Discussion

Each model's parameters applied to both datasets were optimised in Python to provide a fair assessment of each model's performance. Firstly, the key parameters optimised for the decision tree models were max depth (The maximum tree depth), criterion (the measured used to determine split quality), max features (the number of features searched before splitting), and min samples split (the number of samples required for splitting an internal node) [10]. Secondly, random forest models were optimised using similar parameters as those mentioned previously (i.e., max features, max depth and criterion) with the addition of several new parameters such as n_estimators (the number of decision trees used), warm start (using previously fitted models to inform current model), and bootstrap (the process of bootstrapping previously outlined) [10]. Thirdly, kNN models were optimised using four different parameters, which were n_neighbours (the number of data points closest to the centroid), metric (the distance metric), algorithm (the algorithm used to decide the closest neighbours), and weights (the influence given to each neighbour) [10]. Fourthly, MLP underwent optimisation by manipulating the following four parameters: activation (the activation function applied to the hidden layers), solver (the solution provided to optimise weights), hidden layer sizes (the number of neurons in each hidden layer), warm start [10]. Fifthly, Adaboost implemented the previously defined parameters for decision tree models as well as learning rate (the strength of the weights applied at each after each estimator) and n_estimator [10]. Lastly, XGBoost models were optimised via several different parameters, which were booster (the type of boosting method), max depth, n_estimators, objective (the learning objective) [8]. To further understand the specific values of each model for each dataset refer to Appendix 1. Each model was trained on a cross-validated training set using K-fold and then evaluated on a test set. The results of each model on the test set are provided hereafter.

A. NSL-KDD-Dataset

Specifically referring to Table 1, across all the classes the class with the greatest accuracy was U2R (99.10%), whilst the worst accuracy was Normal/Benign (78.48%). The remaining classes (DoS, Prob, R2L) all obtained accuracy scores either approaching or above 90% (91.63%, 94.36%, 89.60%). In contrast to accuracy scores, U2R (45.00%) was the worst-performing class when assessing model precision, however, Normal/Benign was a relatively close second (67.36%). The class with the highest precision was R2L (96.75%) with DoS as a very close second (96.37%) and Prob performing in between the worst and the best two classes (77.64%). Furthermore, class recall for Normal/Benign (97.08%), DoS (78.25%), Prob (66.78%), R2L (9.25%), and U2R (54.50%) achieved rankings in their respective order. Similarly,

Table 1 Decision tree model performance and NSL-KDD-Dataset

Attack	Accuracy (%)	Precision (%)	Recall (%)	F-Score (%)	FPR (%)
Normal	78.48	67.36	97.08	79.54	35.59
DoS (Denial of Service attacks)	91.63	96.37	78.25	86.37	1.51
Prob (Probing attacks)	94.36	77.64	66.78	71.80	2.32
R2L (Root to Local attacks)	89.60	96.75	9.25	16.88	0.04
U2R (User to Root attacks)	99.10	45.00	4.50	8.18	0.05

these rankings are mirrored in F-scores (Prob: 71.80%, R2L: 16.88%, U2R: 8.18%) except for DoS (86.37%) ranking higher than Normal/Benign (79.54%). Lastly, R2L (0.04%), U2R (0.05%), Prob (2.32%) and DoS (1.51%) obtained sound results for FPR, however, Normal/Benign FPR results were substantially higher (35.59%). These results are further highlighted visually by Fig. 1.

Assessing the class level performance measures of the random forest model applied to the NSL-KDD dataset can be located in Table 2. Class accuracy above 90% was achieved by U2R (99.11%), DoS (90.72%), and Prob (94.55%) with R2L reaching similar levels of performance (89.19%), however, Normal/Benign obtained a class accuracy significantly lower (75.67%). Class precision was ranked highest for both DoS (96.17%) and R2L (97.22%), while Normal/Benign (64.38%) and Prob (85.16%) ranked lowest with U2R (33.33%) middling the four classes.

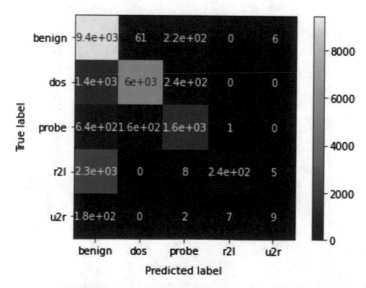

Fig. 1 Confusion matrix highlighting the classification of attacks for the decision tree model

Table 2 Random forest model performance and NSL-KDD-Dataset

Attack	Accuracy (%)	Precision (%)	Recall (%)	F-Score (%)	FPR (%)
Normal	75.67	64.38	97.41	77.52	40.79
DoS	90.72	96.17	75.62	84.66	1.54
Prob	94.55	85.16	59.68	70.18	1.25
R2L	89.19	97.22	5.43	10.30	0.02
U2R	99.11	33.33	0.50	0.99	0.00

Normal/Benign, DoS, Prob, R2L, and U2R achieved respective rankings for class recall (97.41%, 75.62%, 59.68%, 5.43%, 0.50%), whilst inverse rankings were obtained for FPR (40.79%, 1.54%, 1.25%, 0.02%, 0.00%). Similar rankings were achieved for class F-scores for Prob (70.18%), R2L (10.30%), and U2R (0.99%), however, DoS F-score (84.66%) was superior to Normal/Benign (77.52%). A confusion matrix supporting these results can be seen in Fig. 2.

Class level metrics assess the performance of the kNN model applied to the NSL-KDD dataset can be found in Table 3. U2R (99.95%), DoS (91.89%) and Prob (95.41%) were among the highest-ranking classes with each achieving an accuracy score of higher than 90%, while R2L (89.07%) and Normal/Benign (77.83%) both marginally to moderately worse. As for class precision, DoS (96.43%) and Prob (86.91%) again achieved among the highest scores with the addition of R2L (78.84%), whilst U2R (19.67%) and Normal/Benign (66.64%) class precision was among the lowest scores. Regarding class recall, Normal/Benign (97.18%) obtained the highest score followed by DoS (78.97%), Prob (67.44%), U2R (6.00%) and lastly,

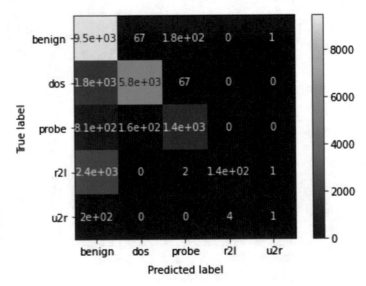

Fig. 2 Confusion matrix highlighting the classification of attacks for the random forest model

Table 3 kNN model performance and NSL-KDD-Dataset

Attack	Accuracy (%)	Precision (%)	Recall (%)	F-score (%)	FPR (%)
Normal	77.83	66.64	97.18	79.06	36.81
DoS	91.89	96.43	78.97	86.83	1.50
Prob	95.41	86.91	67.44	75.95	1.22
R2L	89.07	78.84	5.79	10.79	0.20
U2R	99.95	19.67	6.00	9.20	0.22

R2L (5.79%). Assessing class F-scores, one can observe that DoS (86.83%) was the highest-ranking class with Normal/Benign (79.06%) and Prob (75.95%) achieving similar scores, whilst R2L (10.79%) and U2R (9.20%) F-scores were significantly lower. Lastly, the best ranking FPR score was achieved by R2L (0.20%) then followed by U2R (0.22%), Prob (1.22%), DoS (1.50%) and Normal/Benign (36.81%), which was significantly worse. These results were obtained based on values depicted in Fig. 3.

Table 4 outlines the results obtained from the MLP model applied to the NSL-KDD dataset. The accuracy scores attained indicate that U2R (99.15%) obtained the highest-class accuracy, while DoS (90.58%), Prob (92.18%), and R2L (90.04%) all obtained comparable results with a significant decline in performance from Normal/Benign (76.13%) class accuracy. Furthermore, MLP class precision was highest for R2L (96.34%) and DoS (93.87%), while distinctly lower results were obtained for U2R (60.47%), Normal/Benign (65.81%) and Prob (65.17%). Normal/Benign (92.76%), DoS (77.24%), Prob (58.52%), U2R (13.00%) and R2L

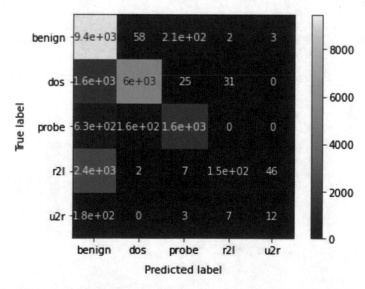

Fig. 3 Confusion matrix highlighting the classification of attacks for the kNN model

Table 4 MLP model performance and NSL-KDD-Dataset

Attack	Accuracy (%)	Precision (%)	Recall (%)	F-score (%)	FPR (%)
Normal	76.13	65.81	92.76	77.00	36.46
DoS	90.58	93.87	77.24	84.75	2.58
Prob	92.18	65.17	58.52	61.67	3.77
R2L	90.04	96.34	13.29	23.35	0.07
U2R	99.15	60.47	13.00	21.40	0.08

(13.29%) achieve respective rankings for class recall. In terms of class F-score, DoS (84.75%) obtain the highest score followed by Normal/Benign (77.00%), Prob (61.67%), R2L (23.35%) and U2R (21.40%). Finally, FPR (36.46%) was highest for Normal/Benign with a significant decline in performance for Prob (3.77%), DoS (2.58%), R2L (0.07%), and U2R (0.08%). These results were attained from the confusion matrix in Fig. 4.

The results of the Adaboost model applied to the NSL-KDD dataset are indicated in Table 5. Referring specifically to class accuracy, one can observe that U2R (99.11%) achieved the greatest accuracy followed closely by Prob (95.54%) and DoS (92.05%). Although R2L (89.29%) and Normal/Benign (76.92%) class accuracy were the lowest ranking accuracies relative to the aforementioned classes the performance was still reasonably high. Transitioning to class precision, R2L (97.63%) and DoS (96.35%) obtained the highest levels with minor differences between the two, whilst Prob (85.15%), Normal/Benign (65.69%) and U2R (50.00%) were all distinctly lower in scoring. As for the class recall, Normal/Benign (97.18%)

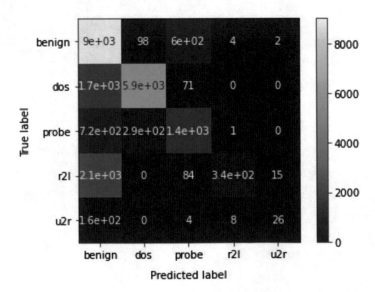

Fig. 4 Confusion matrix highlighting the classification of attacks for the MLP model

Table 5 Adaboost model performance and NSL-KDD Dataset

Attack	Accuracy (%)	Precision (%)	Recall (%)	F-score (%)	FPR (%)
Normal	76.92	65.69	97.18	78.39	38.42
DoS	92.05	96.35	79.53	87.14	1.54
Prob	95.54	85.15	59.64	70.15	1.25
R2L	89.29	97.63	6.41	12.03	0.02
U2R	99.11	50.00	2.00	3.85	0.02

achieve superior results with a continual decline in scores thereafter for DoS (79.53%), Prob (59.64%), R2L (6.41%), and U2R (2.00%). F-scores followed a similar ranking pattern (Prob: 70.15%, R2L: 6.41% and U2R: 3.85%) with exception of DoS (87.14%) ranking higher than Normal/Benign (78.39%). The final measure of class performance was FPR, where rankings were as follows: U2R (0.02%), R2L (0.02%), Prob (1.25%), DoS (1.54%), and Normal/Benign (38.42%). Lastly, Fig. 5: Supports these following results by indicating specifically the classification and misclassification of each class.

The XGBoost class performance metrics on the NSL-KDD dataset are provided in Table 6. Assessing class accuracy one can notice that U2R (99.15%) achieved the highest accuracy with Prob (94.51%), and DoS (92.64%) closely followed, whilst R2L (89.45%) achieved the second-worst class accuracy the accuracy approaches 90% in contrast to the distinct decline in class accuracy for Normal/Benign (78.73%). Regarding class precision, DoS (96.30%) and R2L (99.49%) achieved distinctly higher scores than U2R (83.33%), Prob (80.57%), and Normal/Benign

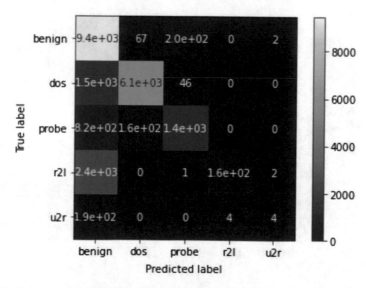

Fig. 5 Confusion matrix highlighting the classification of attacks for the Adaboost model

Table 6 XGBoost model performance and NSL-KDD Dataset

Attack	Accuracy (%)	Precision (%)	Recall (%)	F-score (%)	FPR (%)
Normal	78.73	67.64	97.10	79.73	35.16
DoS	92.64	96.30	81.39	88.22	1.60
Prob	94.51	80.57	64.51	71.65	1.87
R2L	89.45	99.49	7.61	14.15	0.00
U2R	99.15	83.33	5.00	9.43	0.00

(67.64%). In contrast, class recall gradually declines for each subsequent class with Normal/Benign (97.10%) scoring the highest then DoS (81.39%), Prob (64.51%), R2L (7.61%), and U2R (5.00%). Class F-scores varied also with DoS (88.22%) obtaining the highest result followed by Normal/Benign (79.73%), Prob (71.65%), R2L (14.15%), and U2R (9.43%). Finally, FPR was relatively low for all classes (U2R: 0.00%, R2L: 0.00%, Prob: 1.87%, DoS: 1.60%) expect Normal/Benign (35.16%). A confusion matrix indicating the model classifications can be seen visually in Fig. 6.

To assess the performance of each deployed model on the NSL-KDD dataset several overall metrics were used. These metrics as well as the performance of each model are summarised in Table 7. Assessing overall model accuracy, one can observe that each model performed reasonably well with the XGBoost model (77.24%) achieving the highest level of accuracy with a slight decline in performance for the decision tree model (76.59%), the kNN model (76.57%), the Adaboost model (75.96%), the random forest model (74.61%) and the MLP model (74.04%). These

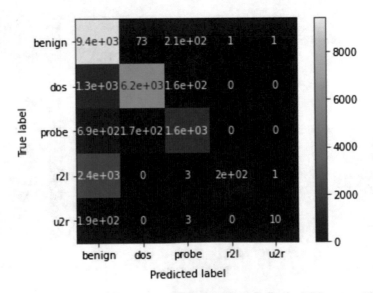

Fig. 6 Confusion matrix highlighting the classification of attacks for the XGBoost model

Table 7 Overall performance metrics for deployed models on the NSL-KDD Dataset

Algorithms	Accuracy (%)	Precision (%)	Recall (%)	F-score (%)	FPR (%)	Fit Time (S)
Decision tree	76.59	76.59	76.59	76.59	5.85	1.31
Random forest	74.61	74.61	74.61	74.61	6.35	21.34
kNN	76.57	76.57	76.57	76.57	5.86	16.28
MLP	74.04	74.04	74.04	74.04	6.49	143.57
Adaboost	75.96	75.96	75.96	75.96	6.01	84.84
XGBoost	77.24	77.24	77.24	77.24	5.69	267.34

results were identical for several other overall metrics (i.e., precision, recall and F-score), which further adds to the consistency of the results obtained. The FPR followed the same rankings of performance as highlighted in the previous metrics with the XGBoost model (5.69%), the decision tree model (5.85%), the kNN model (5.86%), the Adaboost model (6.01%), the random forest model (6.35%) and the MLP model (6.49%). Lastly, a final metric added to assess overall performance, which was not used in class assessments are fit time (i.e., how long does it take to fit the model to the data). Given the following metric, the decision tree model was the fastest (1.31 s) with more than a tenfold increase in fit time for the second fastest, the kNN model (16.28 s). Interestingly, XGBoost (267.34), which was the superior model for all the previous metrics was significantly slower than all remaining models (random forest: 21.34, Adaboost: 84.84, and MLP: 143.57).

B. TON_IoT dataset

The overall results of the deployed models on the TON_IoT dataset are provided in Table 8. The XGBoost (87.14%) and random forest (87.12%) models achieved the highest results across accuracy, precision, recall, and F-score. Adaboost (86.85%) and decision tree (86.48%) models were the third and fourth highest-ranking models in terms of accuracy, precision, recall and F-score, whilst kNN (83.33%) and MLP

Table 8 TON_IoT dataset performance results for deployed algorithms

Algorithms	Accuracy (%)	Precision (%)	Recall (%)	F-score (%)	FPR (%)	Fit time (S)
Decision tree	86.48	86.48	86.48	86.48	13.52	2.31
Random Forest	87.12	87.12	87.12	87.12	12.88	162.41
kNN	83.32	83.32	83.32	83.32	16.68	337.54
MLP	80.43	80.43	80.43	80.43	19.57	217.36
Adaboost	86.85	86.85	86.85	86.85	13.15	55.74
XGBoost	87.14	87.14	87.14	87.14	12.85	331.76

(80.43%) models were the least performative models across these metrics. Furthermore, observing FPR scores indicates that models obtained identical rankings, which were as follows: XGBoost (12.85%), Random Forest (12.88%), Adaboost (13.15%), Decision Tree (13.52%), kNN (16.68%) and MLP (19.57%). Regarding fit time, the decision tree model (2.31 s) demonstrates a clear advantage in fit time with the remaining models ranking as follows Adaboost (55.74 s), random forest (162.41 s), MLP (217.36 s), XGBoost (331.76 s), and kNN (337.54 s). The individual classifications and misclassifications of each can be seen in Fig. 7.

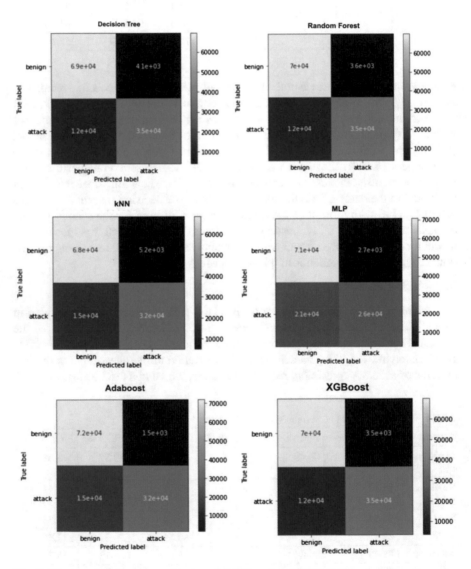

Fig. 7 The overall performance of each model deployed on TON_IoT dataset

6 Conclusion

The following work outlined the performance of several ML models as intrusion-detection models on two benchmark datasets. Using both binary classification and multi-classification tasks a fair assessment of each model could be determined. In both scenarios, XGBoost demonstrated superior classification than the remaining models, however, superior results required considerable differences in processing times. In contrast, decision tree models offer significant gains in processing times, whilst still maintaining relatively high levels of performance across all metrics. The remaining models provided some combination of speed-accuracy trade-off resulting in less than desirable results. When comparing the suitability of ML models, one should always be aware of the notion that no model is necessarily superior in all conditions (i.e., no free lunch) and that adopting an elicit approach is always recommended. Regardless, the decision tree models appear to offer an easy to implement and effective baseline ML model in which alternative models can be compared when using the NSL-KDD or TON_IoT datasets.

Although ML can detect the cyberattacks with very high accuracy, the ML models itself can be vulnerable to cyberattacks, known as adversarial machine learning [11, 12]. In our future work, we will try to focus on such adversarial models. Besides, IDS application can be evaluated considering emerging domains, e.g., Vehicular Network [13], Electricity Grid [14, 15] etc. which will be explored in our future research.

Appendix

NSL-KDD-Dataset - Models deployed.

DecisionTreeClassifier(ccp_alpha = 0.0, class_weight = None, criterion = 'entropy', max_depth = 200, max_features = None, max_leaf_nodes = None, min_impurity_decrease = 0.0, min_impurity_split = None, min_samples_leaf = 1, min_samples_split = 2, min_weight_fraction_leaf = 0.0, presort = 'deprecated', random_state = None, splitter = 'best').

RandomForestClassifier(bootstrap = True, ccp_alpha = 0.0, class_weight = None, criterion = 'entropy', max_depth = 40, max_features = 'sqrt', max_leaf_nodes = None, max_samples = None, min_impurity_decrease = 0.0, min_impurity_split = None, min_samples_leaf = 1, min_samples_split = 2, min_weight_fraction_leaf = 0.0, n_estimators = 200, n_jobs = -1, oob_score = False, random_state = None, verbose = 0, warm_start = True).

KNeighborsClassifier(algorithm = 'auto', leaf_size = 30, metric = 'manhattan', metric_params = None, n_jobs = -1, n_neighbours = 2, p = 2, weights = 'distance').

MLPClassifier(activation = 'relu', alpha = 0.0001, batch_size = 'auto', beta_1 = 0.9, beta_2 = 0.999, early_stopping = False, epsilon = 1e-08, hidden_layer_sizes = (100,), learning_rate = 'constant', learning_rate_init = 0.001, max_fun = 15,000, max_iter = 200, momentum = 0.9, n_iter_no_change = 10, nesterovs_momentum

= True, power_t = 0.5, random_state = None, shuffle = True, solver = 'lbfgs', tol
= 0.0001, validation_fraction = 0.1, verbose = False, warm_start = True).

AdaBoostClassifier(algorithm = 'SAMME.R', base_estimator = Deci-
sionTreeClassifier (ccp_alpha = 0.0, class_weight = None, criterion =
'entropy', max_depth = 200, max_features = None, max_leaf_nodes = None,
min_impurity_decrease = 0.0, min_impurity_split = None, min_samples_leaf =
1, min_samples_split = 2, min_weight_fraction_leaf = 0.0, presort = 'deprecated',
random_state = None, splitter = 'best'), learning_rate = 0.9, n_estimators = 50,
random_state = None).

XGBClassifier(base_score = 0.5, booster = 'gbtree', colsample_bylevel = 1,
colsample_bynode = 1, colsample_bytree = 1, gamma = 0, learning_rate = 0.1,
max_delta_step = 0, max_depth = 7, min_child_weight = 1, missing = None,
n_estimators = 100, n_jobs = -1, nthread = None, objective = 'multi:softprob',
random_state = 0, reg_alpha = 0, reg_lambda = 1, scale_pos_weight = 1, seed =
None, silent = None, subsample = 1, verbosity = 1).

Ton_IoT dataset - Models deployed.

DecisionTreeClassifier(ccp_alpha = 0.0, class_weight = None, criterion =
'gini', max_depth = None, max_features = None, max_leaf_nodes = None,
min_impurity_decrease = 0.0, min_impurity_split = None, min_samples_leaf =
1, min_samples_split = 2, min_weight_fraction_leaf = 0.0, presort = 'deprecated',
random_state = None, splitter = 'best').

RandomForestClassifier(bootstrap = True, ccp_alpha = 0.0, class_weight
= None, criterion = 'gini', max_depth = None, max_features = 'auto',
max_leaf_nodes = None, max_samples = None, min_impurity_decrease = 0.0,
min_impurity_split = None, min_samples_leaf = 1, min_samples_split = 2,
min_weight_fraction_leaf = 0.0, n_estimators = 100, n_jobs = -1, oob_score =
False, random_state = None, verbose = 0, warm_start = False).

KNeighborsClassifier(algorithm = 'auto', leaf_size = 30, metric = 'minkowski',
metric_params = None, n_jobs = -1, n_neighbours = 5, p = 2, weights = 'uniform').

MLPClassifier(activation = 'relu', alpha = 0.0001, batch_size = 'auto', beta_1
= 0.9, beta_2 = 0.999, early_stopping = False, epsilon = 1e-08, hidden_layer_sizes
= (100,), learning_rate = 'constant', learning_rate_init = 0.001, max_fun = 15,000,
max_iter = 200, momentum = 0.9, n_iter_no_change = 10, nesterovs_momentum
= True, power_t = 0.5, random_state = None, shuffle = True, solver = 'adam', tol
= 0.0001, validation_fraction = 0.1, verbose = False, warm_start = False).

AdaBoostClassifier(algorithm = 'SAMME.R', base_estimator = Decision-
TreeClassifier (ccp_alpha = 0.0, class_weight = None, criterion = 'gini', max_depth
= 10, max_features = None, max_leaf_nodes = None, min_impurity_decrease =
0.0, min_impurity_split = None, min_samples_leaf = 1, min_samples_split = 2,
min_weight_fraction_leaf = 0.0, presort = 'deprecated', random_state = None,
splitter = 'best'), learning_rate = 1.0, n_estimators = 50, random_state = None).

XGBClassifier(base_score = 0.5, booster = 'gbtree', colsample_bylevel = 1,
colsample_bynode = 1, colsample_bytree = 1, gamma = 0, learning_rate = 0.1,
max_delta_step = 0, max_depth = 30, min_child_weight = 1, missing = None,
n_estimators = 500, n_jobs = -1, nthread = None, objective = 'reg:squarederror',

random_state = 0, reg_alpha = 0, reg_lambda = 1, scale_pos_weight = 1, seed = None, silent = None, subsample = 1, verbosity = 1).

References

1. D. Meharch, ani on April 20, and 2021, '10 Major Cyber Attacks Witnessed Globally in Q1 2021', Security Boulevard, Apr. 20, 2021. https://securityboulevard.com/2021/04/10-major-cyber-attacks-witnessed-globally-in-q1-2021/ (Accessed 05 May 2021).
2. CIC and ISCX, 'NSL-KDD dataset'. https://www.unb.ca/cic/datasets/nsl.html (accessed May 05, 2021).
3. A. Alsaedi, N. Moustafa, Z. Tari, A. Mahmood, A. Anwar, TON_IoT telemetry dataset: a new generation dataset of IoT and IIoT for data-driven Intrusion detection systems. IEEE Access **8**, 165130–165150 (2020)
4. E. Tabassi, K.J. Burns, M. Hadjimichael, A.D. Molina-Markham, J.T. Sexton, A taxonomy and terminology of adversarial machine learning', preprint, Oct. 2019. https://doi.org/10.6028/NIST.IR.8269-draft
5. M. Tavallaee, E. Bagheri, W. Lu, A.A. Ghorbani, A detailed analysis of the KDD CUP 99 data set (2009), pp. 1–6
6. A. Géron, *Hands-on machine learning with Scikit-Learn, Keras, and TensorFlow: Concepts, tools, and techniques to build intelligent systems* (O'Reilly Media, 2019)
7. J. Friedman, T. Hastie, R. Tibshirani, *The Elements of Statistical Learning*. Springer series in statistics, vol. 1, no. 10 (New York, 2001)
8. T. Chen, C. Guestrin, Xgboost: a scalable tree boosting system (2016), pp. 785–794
9. S. Haghighi, M. Jasemi, S. Hessabi, A. Zolanvari, PyCM: Multiclass confusion matrix library in Python. J. Open Sour. Softw. **3**(25), 729 (2018)
10. F. Pedregosa et al., Scikit-learn: Machine learning in Python. J. Mach. Learn. Res. **12**, 2825–2830 (2011)
11. M.A. Husnoo, A. Anwar, Do not get fooled: defense against the one-pixel attack to protect IoT-enabled deep learning systems. Ad Hoc Netw. **122** (2021)
12. M. Billah, A. Anwar, Z. Rahman, S.M. Galib, Bi-level poisoning attack model and counter-measure for appliance consumption data of smart homes. Energies (2021)
13. S.T. Mehedi, A. Anwar, Z. Rahman, K. Ahmed, Deep transfer learning based intrusion detection system for electric vehicular networks. Sensors (2021)
14. A. Anwar, A.N. Mahmood, Z. Shah, A data-driven approach to distinguish cyber-attacks from physical faults in a smart grid, in Proceedings of the 24th ACM International on Conference on Information and Knowledge Management (CIKM '15). Association for Computing Machinery, New York (2015), pp. 1811–1814
15. B.M.R. Amin, M.J. Hossain, A. Anwar, S. Zaman, Cyber attacks and faults discrimination in intelligent electronic device-based energy management systems. Electronics (2021)

Chapter 6
Artificial Intelligence for Cyber Security: Performance Analysis of Network Intrusion Detection

Shahriar Usman Khan, Fariha Eusufzai, Md. Azharuddin Redwan, Mohiuddin Ahmed, and Saifur Rahman Sabuj

Abstract Cybersecurity has become major progress in the digital era. Contraction is an important component of the cyber analysts' management of information technology, as several government organizations and commercial enterprises are moving to dispersed systems. A cyber security analyst is most importantly responsible for protecting the network against damage. Attacks on networks are becoming more complex and sophisticated every day. The number of connected workplaces leads to heavy traffic, more security attack vectors, security breaches and raises more issues than the cyber area can handle by using human intervention while there is not enough sizable automation. Network Intrusion, thus, becomes the biggest concern of this generation. Intrusion Detection and Prevention System (IDPS) has become a vital complement to almost all organizations' security infrastructure. This chapter includes three network monitoring tools concentrating on the immediate impact and output of cyber assaults on the network. This helps to better comprehend the many directions in the area of network monitoring and cybersecurity research. In general, this chapter aims to study the different network monitoring techniques that can be used in support of various servers and network devices to better understands the effect of cyber-attacks on the network and monitor it through tools such as cacti, weather-map and smokeping. It also sheds light on techniques like artificial intelligence, machine learning, neural networks, fuzzy logic, next-generation firewall and how they can be coupled with Intrusion Detection System (IDS) to detect attacks on private networks.

S. U. Khan (✉)
Institute of Information Technology (IIT), Jahangirnagar University, Dhaka, Bangladesh

F. Eusufzai · Md. Azharuddin Redwan
Department of Electrical and Electronics Engineering, BRAC University, Dhaka, Bangladesh

M. Ahmed
School of Science, Edith Cowan University, Joondalup, Australia
e-mail: mohiuddin.ahmed@ecu.edu.au

S. R. Sabuj
Department of Electronic Engineering, Hanbat National University, Daejeon, South Korea
e-mail: s.r.sabuj@ieee.org

© The Author(s), under exclusive license to Springer Nature Switzerland AG 2022
M. Ahmed et al. (eds.), *Explainable Artificial Intelligence for Cyber Security*,
Studies in Computational Intelligence 1025,
https://doi.org/10.1007/978-3-030-96630-0_6

Keywords IDPS · Network monitoring tools · Cyber-attacks monitoring · Cybersecurity · Next generation firewall

1 Introduction

The defense of computer systems and networks against the exposure of information theft or loss of hardware, software or electronic data as well as the interruption or error of functionality provided by them is Cybersecurity or Information Technology (IT) security. Wireless technology is one of digital technology's most diverse domains. The area is highly important as computing networks, web and wireless network standards are becoming increasingly dependent on them such as Bluetooth, Wi-Fi and the proliferation of "intelligent" devices, including tablets, televisions and the numerous features that make up the "Internet of Things." Because of its international and technological complexity, cybersecurity is still one of the biggest problems of today's modern world. New terms such as cyber-attack and cyber warfare have contributed to an increasingly digital world dependent on information and computer networks in both the corporate and government sectors including the military. In times of conflict, certain technologies have been identified as priorities as vital infrastructures which can seriously destabilize a nation. The risk of cybersecurity is growing due to global networking and the use of cloud technology, to store classified information and personal data. The fact that cloud systems are not well configured for more advanced cybercriminals means that the risk of active cyber-attacks or data breaches is rising.

The importance of cybersecurity, as it defends against misuse and damage of all data types. These include classified data, confidential personal health information (PHIs), personal data, intangible assets, knowledge structures of the government and the industry. Artificial intelligence has been incorporated into defense networks to reduce the ever-growing cyber-security risks that multinational companies experience. There is a rising study in the subject of cyber security which concentrates on the knowledge of safety analysts' work processes [5]. Security in data communication networks is designed to safeguard the space that exists inside a network or emerges as a result of the network administrator's lack of knowledge in managing data communication network protection [12]. Cyber-attacks are part of our existence which affects all countries involved in the industry. Given the factual deduction of cyber defense, the safety of their various cyber-IT ecosystems against threats such as theft, vandalism, vulnerabilities and spying are important for all businesses and organizational sectors. Although several technical developments in information and network security have aided experts in enhanced observation and threat detection, the operations they undertake cannot be entirely computerized. The Intrusion Detection System (IDS) is a technology that security analysts depend significantly on [5].

Cloud Computation provides individuals and enterprises with on-demand, comparatively cheap, customizable and hassle-free IT infrastructure making it a popular service due to the advancement of the internet. Cloud computing utilizes the distributed method to make the services available which highly increases the security breaches and integrity of the data. In an era where cheap and mobile computational devices are abundant, various cyberattacks can be easily carried out which may greatly hamper the integrity and security of data [20]. Incidents involving Amazon's Elastic Compute Cloud (EC2) cloud infrastructure and the powerful attacks from EC2 is indeed raising serious concerns in the community of system administrators and security experts. An IDS is a solution to these security issues. An IDS is a software or hardware device that can analyze internal and external network activities of cloud infrastructure, it can sense an abnormal behavior, which may pose an attack on the system, and automatically notify the network or system administrator [24]. Carefully tailored IDS systems can effectively help administrators prevent cyberinfrastructure from cyber-attacks. There are different types of IDS based on different methods of action cybersecurity [24].

This Chapter contains three network surveillance techniques focused on the instantaneous influence and output of cyberattacks. This enables the numerous paths within network observation and cybersecurity development to be better understood. The main findings of this chapter are summarized below. It also provides information on several strategies and how intrusions on private networks may be detected by IDS.

1.   We used a centos Linux server to configure network monitoring tools to monitor bandwidth, traffic load and packet loss of the network
2.   We generated Denial of Service (DoS) and Distributed Denial of Service (DDoS) attacks in Kali Linux to show the effect of the cyberattacks on the network.
3.   This system model shows how we can create cyber-attacks inside a network and shows the output of those cyber-attacks through network monitoring tools using different network protocols for instance Simple Network Management Protocol (SNMP), Network Time Protocol (NTP) etc.

The chapter continues following the description. Literature reviews and the background of firewall and various types of attacks, including the definitions of the server is demonstrated in Sect. 1. Section 2 illustrates the entire network architecture in detail. In Sect. 3, we have explained the simulation results which we have obtained through every tool. Finally, we have concluded our chapter via Sect. 4.

2   Literature Review of Existing Works

Reviewing the literature is an important aspect that provides a useful framework for information advancement, which initiates discovering areas where there is a need for development. In AI-based and Cloud-based Cyber-Security many professionals have proposed different mechanisms and models to establish their thoughts, motivating us to research more on this topic.

2.1   Review of AI-Based Cyber-Security

AI systems may help identify and take appropriate measures to resolve such conditions, classify incidents and risks and finally relieve engineers from routine tasks. Alberto Perez Veiga talked about the necessity for cybersecurity strategies to develop and the use of Artificial Intelligence (AI) to aid solve certain issues [3], also to finalize his analysis on what is the anticipated future of AI's use on network security, he presented an elevated summary of certain highly advanced AI network safety tactics. Similarly, Radha R [4] and the authors [7–11], not only described the relevance of AI, along with cybersecurity but also illustrated its huge necessity against cyber-attacks. Besides, Tyugu, Enn [6] has explained a similar concept along with the possibility of neural networks that can be used to combat different cyber threats. Furthermore, authors [5] explained the findings relate to cognitive processes and discussed their significance for developing cyber security. Moreover, the authors [5] examined the underlying cognitive processes involved in cyber threat detection with a particular emphasis on comprehending the interaction of technical experience and cognitive skills.

An IDS is a software program or computer that detects and notifies the program or network operations in the event of policy breaches or malicious behaviors. This study [13] suggested a new method called outlier detection, where anomaly data collection is evaluated by the Neighborhood Outlier Factor (NOF). Moreover, the trained model here serves the purpose of large datasets with a distributed database framework to enhance intrusion detection system efficiency. Additionally, the authors of this paper [14] briefly observed the foundations of AI in the area of intrusion detection and presented a summary of relevant work to demonstrate the sophistication of the topic, along with some of the most common AI algorithms. Similarly, the authors of this paper [15] presented the various AI techniques that can be used to promote intrusion detection systems in the area of intrusion, techniques such as ML, neural network and Fuzzy Logic and the way they can be linked with the IDS for the detection and use of threats on secure networks. Furthermore, the authors [16] have concentrated on a detailed idea of ML and DL along with a brief idea of machine and deep learning and analyzed a Network-based Intrusion Detection System (NIDS) diligently. Likewise, the authors [17, 18] have portrayed valuable information on the implementations, necessity and significance of AI-related IDS techniques. Nevertheless, there is one more term similar to this concept. Next-Generation Intrusion Prevention System (NGIPS) is a web traffic monitoring system to recognize suspected behavior and early intrusion prevention which can cause network failure as it should have been. The authors of this paper [12] focused on the study and assessment of NGIPS, the defense of network against intrusion schemes that use firewall vulnerability, i.e., HTTP port. Similarly, the author of this paper [19] included an outline of the current security mechanisms and examined their deficiencies and described the Next-generation IDS specifications.

The firewall, which safeguards a network by keeping entry points, is another protection function that is under consideration. A firewall is a network security mechanism intended to block non-authorized network connectivity. In this article [31], the authors suggested adding firewalls to artificial intelligence, which in turn would allow firewalls to manipulate traffic behaviors by using incoming traffic and by generating certain rules on certain extraordinary packages with themselves; no default activities are therefore necessary. Likewise, Abie, Habtamu [32] and the authors [33] discussed all the essential facets of firewall requirement and the importance of firewall in network security and firewall implementation in software/hardware or both respectively. Besides this, the authors of this paper [35, 36] also addressed the firewall vulnerability by type and different types of firewalls, and some of the present firewall mitigation approaches were also stated as a technique that enables attackers to acquire more specific data on the vulnerability of firewalls, to take advantage of these loopholes respectively. The authors of this paper [34] proposed a model hybrid learning Web application firewall (WAF) model, using Signature-based detection (SBD) and Anomaly-based detection (ABD), to avoid web-based attacks. Web-based web attack detection is conducted using SBD, while HTTP error request detection is performed using ABD and ABD-based learning is carried out using artificial neural networks (ANN).

In this model, various AI approaches focusing on the creation of IDS were examined and better comprehend the different avenues for IDS analysis. In modern computing infrastructures, network intrusion detection systems (NIDS) are important for monitoring and identifying unwanted and harmful network activity. The authors [1] outline a non-signature-dependent fraudulent shellcode identification system based on artificial neural networks. Moreover, findings showed that this innovative approach to identification can recognize shellcode with extraordinarily high precision and small amounts of wrongful convictions. A firewall is a software or hardware program that filters traffic via a preset framework. According to [2], along with a neural network, the firewall was also incorporated and it has been used in a multi-layer perceptron to check different packets while transporting pieces of information.

Alongside AI-based cybersecurity, another concept of optical networks was encouraged. In this paper the authors [37] addressed the criteria for implementing optical network solutions and outlined strategies like AI-based QoT and Impairment Modelling, promoting the concept of optical performance monitoring OPM controllers that can modify physical layer configurations and identify inconsistencies.

2.2 Review of Cloud-Based Cyber-Security

With the increased implementation of the Internet and its services, Cloud Computing has become more popular as many organizations started investing in it. Cloud

Computing allows individuals and organizations to access all the necessary IT infrastructures with lesser cost, relatively less time and without the need for having a dedicated team in the organization itself. As more people are starting to take advantage of the internet, an increase in cyber-attacks to exploit the data stored in the cloud occurs which poses a huge barrier in the advancement of Cloud Computing. Al-Jaser [26] in his survey has found that at least 20% of respondents did not know any information about cloud security. He emphasizes the education of the employee regarding cyber-security as it creates a vulnerability of the data. On the other hand, Onyema, E.M. et al. [39] have recommended the incorporation of cloud computing in educational institutions and providing necessary knowledge and skills to the students for better understanding and awareness of cybersecurity in Cloud Computing. Savita et al. [25] and authors [28, 29, 40] have elaborated on the classification of security threats to the cloud infrastructure stating that cyber-attacks can compromise Integrity, Confidentiality and Availability of cloud services. Moreover, the authors provided some prevailing solutions for various security threats.

To combat the ever-evolving cyber-attacks and various other threats of cloud computing many authors have proposed new and improved methods of IDS/IPS. Mazzariello et al. [20] have argued that traditional methods of deploying IDS systems at the edge of the defended systems do not efficiently ensure the security and integrity of the on-demand/pay-per-system cloud-based enterprises. Hence, the authors have proposed to deploy lightweight IDS on all the devices hosting cloud services for the enterprise. Confidentiality and information integrity have become a major topic for individuals and corporate corporations as it hinders the usual business activities. A strong and precise IDS can help secure the confidentiality and integrity of the resources. The authors of the paper [21] have proposed and demonstrated a technique of intrusion detection using SVM, employing an NSL-KDD datasheet to achieve accuracy. Moreover, in their research Chowdhury et al. [42] have proposed an IDS, combining the information gain and decision tree-based algorithm. They claim that the proposed method not only reduces the features to a significant level but also has a high detection rate in both binary and multiclass categories. To add to that, authors [24, 27] have also proposed various IDS approaches including Network-based Intrusion Detection System (NIDS) and Host-based Intrusion Detection System (HIDS). In elaborate research on Mobile Cloud Computing (MCC) and Cloud Computing (CC) authors [28] have presented a survey of an IDS based on Computational Intelligence (CI). The authors have argued that there is a lack of proper IDS for MCC and proposed that well-tuned CI-based IDSs can reach the goal of higher accuracy. In another study, Moorthy et al. [45] have found that Cloud-based IDS were able to detect 80% of random sets of cloud attacks. By adding background traffic retrieved from DARPA, IDS was able to detect the same percentage of attacks and no false-positive cybersecurity alarm was a massive problem raised while filtering background traffic. The efficiency of the proposed cloud IDS was determined by injecting attacks with increased background traffic.

2.3   Network Monitoring Tool (NMT)

A network monitoring tool (NMT) helps the user to analyze the traffic a system has encountered at the time of connection with a computer. NMT provides insightful information on the traffic based on the observation. NMT allows network administrators to observe networks irrespective of their size by providing precise and accurate data on network connectivity and performance. Moreover, it allows one to manage and perform certain tasks on the network making it robust and efficient.

There are different types of network monitoring tools available, but we have employed three main tools – Cacti, Weathermap and Smokeping. A description of the tools is given below:

Cacti

Cacti is a complete interface of RRDTool, it stores all the information needed to create a chart and fills the data in the MySQL database. This interface is fully supported by PHP. In addition to this ability to maintain charts, data sources, and loop files in the database, cactus also handles data collection. For those used to create flow graphs using MRTG, there is also SNMP support.

Advantages of Cacti

Measuring availability, load, errors etc. with the history

- Cacti can display a system's routers, switch interfaces and their traffic, including all error traffic.
- Cacti can measure drive capacity, CPU load (network h/w and servers) and it can respond to incidents and send notifications depending on the ranges of the drives

Graphical capability

i.   Facilitates the use of the functionality of RRDGraph to define graphics and automate the display.
ii. Capability of organizing information in the hierarchical tree structure.

Data collection

i.   Supports SNMP, including use of PHP-SNMP or net-SNMP
ii. Updates data sources via SNMP or defines scripts to capture necessary data
iii. *Cactid* uses multithreading to implement SNMP routines in C

Template

i.   Creates templates to reuse graphics Definition, Data and Device Source

Cacti plugin architecture

i.   Extend Cacti functionality. There are many, many plugins available. Part of the default Cacti installation of Ubuntu 12 and later.

User management

i.   Manage users locally or via LDAP
ii. Assign detailed authorization levels by the user or user group

Network Weathermap

Weathermap is an open-source network visualization tool used to retrieve the data system already has and can show an overview of the target network's activities in the form of a map.

Features:

i.   Data collection is achieved using plugins. Plugins are provided for RRDtool, MRTG (RRD and legacy record formats), tab-delimited text files, SNMP, ping, external scripts, and Cacti-specific data. The RRDtool plugin means that you can access data from a host of open source monitoring tools, including Cacti, Cricket, Zenoss, MRTG, Routers2, Munin, and more. Supports other sources through plugins or external scripts.
ii. Provides detailed documentation and an interactive editor, allowing users to create their maps as easily as possible.
iii. It has a strong Cacti integration, using the Cacti plug-in architecture to provide a map access control and management user interface using Cacti's existing user database. Additional Data Sources plugins allow efficient access to data directly from Cacti poller, as well as data from other Cacti plugins such as THold and DSStats.

Smokeping

Smokeping is a latency measurement tool. It sends test packets over the network and measures the time it takes them to travel from one place to another and back again.

For each measurement cycle, smokeping sends several packets. It then sorts the different round-trip times and chooses the average time (i.e., time in between). This means that when there are 10-time values, the number 5 is selected and drawn. Other values are plotted as successive lighter greys in the background (smoke).

Sometimes a test packet is sent but never returned. This is called a package. The colour of the middle line changes depending on the number of lost packets.

All of this information combined indicates the third-party health of the network. For example, packet loss should not occur. It could mean that a device in the middle of the link is overloaded or that the router configuration somewhere is wrong.

Large fluctuations in RTT (Round Trip Time) values also indicate that the network is overloaded. It appears on the graph as smoke; the more smoke there is, the more vibration there is.

Unlimited Smokeping only checks the round-trip time of packages. It can also perform some remote tasks ("explorations"), such as downloading a web page. This will provide an aggregate "picture" of web server availability and network health.

2.4   Background of Firewall and Various Types of Attacks

Firewall

Firewalls, as the initial line of protection, are becoming increasingly important in network security as communications systems expand. A firewall is one of the most powerful security guards often employed in any network [35]. A firewall is a network security system that regulates and monitors the whole of inbound and outbound network traffic under sophisticated and specified security standards. All packets that enter or leave the network have to pass the firewall that checks each message and rejects those which do not fulfil the security requirements defined [31]. Both software and hardware devices can be set up for Firewall [32]. These sorts of acts require the following criteria: protecting the resources, verifying and functioning as intermediate access, network management and control traffic, registration and reporting of events (Fig. 1).

The basic mechanisms of firewall

A firewall can be constructed from a basic vulnerability scanner to a pairing of many firewalls in different sorts of settings [33]. The basic operations of a firewall are depicted below:

1. Firewall investigates all traffic routed between the network entities or between two separate hosts to verify if specific connecting or passing messages requirements are fulfilled.
2. After that Firewall will filter both inbound and outbound traffic and maintain traces of the outgoing requests on hosts or networks. The firewall allows the inbound traffic if all the recorded network information match precisely.
3. It can also filter packets depending on the addresses and port number of the origin and recipient. For example, "Source address 172.18.1.1 is allowed to reach destination 172.18.2.1 over port 22."

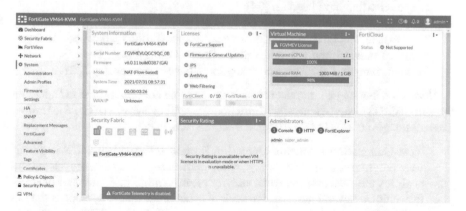

Fig. 1 Physical interface of FortiGate firewall

Types of Firewalls

The firewall for software is used for personal PCs and larger networks are used for hardware firewalls [33]. The following are several types of firewalls and their characteristics:

1. **Packet filtering Firewall**: Firewalls are typically used for packet filtering on the wireless networks that link the internal system. It can only be implemented on the Network Layer of the OSI Model [33].
2. **Stateful Packet Inspection Firewall**: It can often be interpreted as 'Dynamic Packet Filters.' A stately firewall for inspection enables or prevents state-, port and protocol-based traffic. This firewall works on the OSI model's Layer 4 (Transportation Layer) [33].
3. **Application Layer Firewall**: Any OSI layer, up to the Application Layer, can be inspected and filtered by an application layer firewall. It may restrict particular material and detect when certain applications and protocols (such as HTTP and FTP) are being abused. Application Layer Firewalls are also referred to as Proxy Firewalls and Gateway Firewalls [33].
4. **Circuit Level Firewalls**: Circuit-level gateways operate at the OSI model's Session Layer, or as a "shim-layer" bridging the program and transport layers of the TCP/IP stack. To evaluate whether the desired link is authentic, a circuit-level gateway watches TCP handshaking between packets from trustworthy clients or servers to insecure hosts and vice versa [33].
5. **Stateful Multilayer Firewalls**: Stateful multilayer firewalls integrate features from different types of firewalls. They filter packets at the network layer, check whether session packets are genuine, and analyze packet features at the application layer [33].
6. **Web Application Firewall**: WAF is a form of firewall that filters, monitors, and stops HTTP traffic on a web application server's application layer. It is a piece of software or hardware that can listen in on the communication between a web server and a web browser. It identifies web-based attacks depending on the policies that have been set [34].
7. **Next-Generation Firewalls (NGF)**: Deep packet inspection is provided by NGF to detect and redirect conceivably hazardous traffic. It allows for the separation of internal and external traffic. Check Point, Cisco, Dell, Barracuda, Fortinet, Juniper, Palo Alto, and many more businesses develop NGF.

   NGF has a few major components such as - Next-Generation Intrusion Detection System (NGIDS), Next-Generation Intrusion Prevention System (NGIPS), Unified Threat Management (UTM) and Threat Focused NGFW.

   i. **Next-Generation Intrusion Detection System (NGIDS)**: An IDS is a software program or device that analyzes and informs the management systems and analyzes the system or network activity against traffic infractions or destructive actions. Several systems can seek to avoid an intrusion attempt, although neither a monitoring system is necessary nor anticipated. IDPS has become a vital complement to almost any organization's network security. Different approaches may be employed for intrusion detection; however, each method is exclusive to a certain method. The fundamental objective of an intrusion detection system is to efficiently detect assaults. It is also vital at an early stage to detect assaults to limit their consequences [13].

ii. **Next-Generation Intrusion Prevention System (NGIPS)**: NGIPS is a network security system that monitors internet congestion, identifies dubious actions, and prevents intrusion and events at an early stage that might cause the network to fail to function as it should be. This next generation of IPS not only helps control the risk but also enables the IT security developers to perform on the susceptibility of effort recovery or attack and act more successfully if the incidence occurs [12]. The IPS notifies any occurrences to system authorities and acts as a prevention to avoid further assaults such as close access points and setting up firewalls to discover organizational security concerns, IPS technologies may also be used to prevent workers and network visitors from breaching the restrictions included in these standards.

iii. **UTM (Unified Threat Management)**: UTM is a security information method where numerous security functions are provided by a single hardware or software installation. UTM firewalls defend incoming and outgoing traffic from many risks and forms of attacks. On the same centralized platform websites & web content might be blocked & supervised. UTM firewall development increased total network security and lowered complexity and expense, which made endpoint protection solutions available to little companies or other SMBs, where it would be hard to.acquire expensive IT personnel.

Cyber-attacks and their Cross-site third-party variation

Cyber-attacks are attacks that are carried out on a computer or on a computer network to steal, change or demolish crucial and valuable data stored in it. These attacks are executed either by a person or a process gaining unapproved access. Malicious codes are deployed to disrupt the system's integrity and gain or make valuable information unavailable to the user.

Types of Cyber-attack

A. Denial-of-service (DoS) and distributed denial-of-service (DDoS) attacks: Denial Denial-of-service (DoS) attack overruns a system by sending enormous amounts of traffic to disrupt, even paralyzing the whole system and making it unavailable for the user. Similar to the technique, a Distributed Denial-of-service (DDoS) attack is executed when the attacker illegally gains control over several machines or systems to launch an attack. The attacker is not directly benefited from these attacks, but the significant damage to the host is done as a large amount of traffic can cause the host financial loss if the system runs on renting cloud servers. Moreover, disruption of service for a while can disrupt the normal functionality of the targeting system.

Types of DoS and DDoS attacks: TCP SYN Flood attack, ping flood attack, teardrop attack and smurf attack are the different types of DoS and DDoS attacks.

Fig. 2 IPS interface and options in firewall

TCP SYN Flood DDoS attack: It is a type of DDoS attack that takes advantage of a part of a usual TCP three-way handshake to paralyze the target system. To execute such an attack, the attacker sends TCP connection requests quicker than the capability of the target machine's ability to process those requests, resulting in network overflow (Fig. 2).

Attack methodology

When a client and server establish a normal TCP "three-way handshake," the process looks like this:

i.   Client requests a connection by sending a Cross-site SYN (synchronize) message to the server.
ii.   Server acknowledges by sending SYN-ACK (synchronize-acknowledge) message back to the client.
iii. Client responds with an ACK (acknowledge) message, and the connection is established.

In a TCP SYN Flood DDoS attack, the attacker continuously sends SYN packets to all the ports of the targeted server, often using a fake IP address.

During this period, the server cannot close the connection by sending an RST packet and the connection remains open. Before the connection times out, another SYN packet will arrive. This means that more and more connections are half-open; in fact, TCP SYN Flood DDoS attacks are also called "half-open Cross-site" attacks. Over time, as the server connection overflow table fills up, services to legitimate clients will be denied and the server may even crash or crash.

Although the aforementioned "classic" TCP SYN Flood DDoS attempts to exhaust network ports, SYN packets can also be used in DDoS attacks, trying to block systems' pipes with bogus packets to create network saturation. Here, the type of package is not important.

Ping flood attack: Ping flood DDoS, also known as ICMP flood, is a typical DDoS technique in which an attacker overwhelms a victim's computer with ICMP echo requests, also known as pings, to bring it down.

The attack entails sending a huge number of request packets to the victim's system, knowing that the network will respond with an equivalent amount of reply packets.

Attack Methodology: Generally, a ping request is utilized to test the connectivity of two computers by measuring the time consumed from sending an ICMP echo request to receiving an ICMP echo reply. However, during a ping flood DDoS attack, they are deployed to overload the target network with data packets.

The execution of a ping flood DDoS depends on whether the attacker knows the IP address of his target. Therefore, attacks can be divided into three categories based on the method of resolving IP addresses in the target system.

i.   **Targeted local public ping flood** targets a single computer on the local network. For this attack to be carried out, the attacker needs physical access to the computer to discover its IP address. A successful attack will cause the target computer to shut down.

ii. **Router disclosed ping flood** targets the router to disrupt communication between computers on the network. It is based on the attacker knowing the internal IP address of the local router. A successful attack will cause all computers connected to the router to shut down.

iii. **Blind ping flood** involves using an external program to disclose the IP address of the target computer or router before executing the attack.

In a continuous ping flood, the attacking computer must have access to more bandwidth than the victim. This limits the ability to perform DoS attacks, especially for large networks.

In addition, Distributed Denial of Service (DDoS) attacks performed using botnets are more likely to maintain ping flooding and flood target resources.

It is quite difficult to prevent DoS attacks as it is tricky to differentiate a real request from a malicious one. Due to the severity and the magnitude of disruption these attacks can create, we have enforced TCP SYN flood attack and Ping flood attack of the DoS/DDoS attack in our system to observe how precise the Network Monitoring Tools are in determining monitoring these attacks.

Description of other Cyber Attacks follows.

B. SQL injection attacks: SQL injection attacks are when an attacker injects a malicious SQL script[1] into a Web application to access a database stored on the server. A common method for hackers to do this is to inject hidden SQL[2] queries into web forms (such as login forms). Generally, when a user enters their information in a form and presses the "Login" button, an SQL query will be sent to the database to request the user's information. However, when hackers inject malicious SQL queries, they can request various data from the database. At that time, hackers can

---

[1] The SQL script contains a set of SQL commands saved as a single file.

[2] The SQL query is a request for data or information from a database.

easily view, change or delete data, and may paralyze the entire system. Because most web applications store databases on their servers, these applications become attractive targets for SQL injection, which can lead to the disclosure of confidential information.

C. Cross-site third-party scripting (XSS) attack: When a website has a vulnerability that allows script injection, a cross-site scripting attack occurs. Attackers exploit these vulnerabilities and inject malicious JavaScript into the website's database. When the user subsequently requests this data, the user's web browser will execute malicious JavaScript. This will allow an attacker to steal the browser's cookies to hijack session (see Footnote 1). Hackers can use session information to exploit other vulnerabilities, possibly obtain information from the network, and control the user's computer. This is especially important for corporate environments, as XSS attacks can compromise the entire network (see Footnote 1).

Every time a user logs in to a website account, the browser records it as a session and stores it as a cookie. Session hijacking is when an attacker steals the browser's cookies and tricks the browser into thinking that it is the same as the original session, allowing the attacker to gain full access to the user's account.

D. Broque Force Strength Attack: Brute Force Attack, one of the easiest ways to web attacks, sometimes called password attacks. Attackers repeatedly try different combinations of usernames and passwords until the attacker logs into the victim's account. For example, if we take 52 characters (uppercase and tiny) and a third-party standard 10-digit password. A possible combination of $62^8 = 2.1834011 \times 10^{14}$ is performed. When a hacker controls multiple computers or develops a powerful software-based computing engine, the task becomes easy for the attacker.

E. Zero-day attack: Zero-day attack exploits published software flaws before the software update fixes. If the day of the software update is considered the "first day", then the attack that occurred before the update is called the "zeroth day", making it a zero-day attack. Because these attacks target deficiencies that have not yet been repaired, they may be difficult to repair in the short term. It can be compared with new viruses without a vaccine or treatment.

Many zero-day attacks target Microsoft Windows immediately after the software update is released, exposing those who do not update immediately.

F. Man in the middle attack (MitM): A man in the middle attack occurs when a third party intervenes in the communication between the client and the server. The third party pretends to be client and server and gains access to information between them. This type of attack allows threat actors to obtain, send and receive data intended to be provided to others. MITM attacks abuse the real-time operation of transactions, communications, or the exchange of other information. The different types of man-in-the-middle attacks include session hijacking, IP spoofing, and response. You can configure an intrusion detection system to prevent man-in-the-middle attacks. If someone tries to hijack the transmission from the network, it helps to generate an alert immediately. Virtual private networks can also be used to prevent man-in-the-middle attacks. This helps create additional layers of security when accessing the confidential layer of the company over Wi-Fi.

3 System Model

This system model is mainly designed to create Dos and DDoS attacks in any network and generate the output of those attacks in three network monitoring tools cacti, weather map and smokeping. Moreover, this system model showed how cyberattacks are created inside a network and how we can identify those attacks by monitoring traffic loads, interface status and packet loss percentage in these three monitoring tools. This system model was created primarily to observe the network traffic and bandwidth status and the degradation of each component inside the framework. This system model is implemented in gns3 software alongside with virtualization tool VMware workstation software that helps cloud 1, cloud 2 and cloud 3 adapters to connect with physical machines and access the system model devices and monitoring tools from the physical machine (Fig. 3).

In this topology, we used one Centos Linux Server, one Kali-Linux server, two cisco routers, one Juniper router, one Mikrotik router and one FortiGate next-generation firewall.

We configured a total of 3 network monitoring tools (cacti, weather map, smokeping) in the Centos-Linux server to monitor our network and we used the kali Linux server to generate cyberattacks such as Dos and DDoS attacks.

In this topology, the following IP addresses were assigned for Devices:

Centos-Linux

Ethernet 0: 192.168.15.0/24.
Ethernet 1: 192.168.89.145/24.

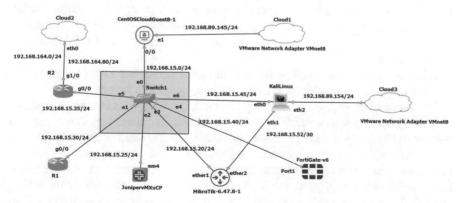

Fig. 3 Network architecture

Kali-Linux

Ethernet 0: 192.168.15.45/24.
Ethernet 1: 192.168.15.54/24.
Ethernet 2: 192.168.89.154 /24.

Cisco- R1

Gigabitethernet 0/0: 192.168.15.30/24.

Cisco- R2

Gigabitethernet 0/0: 192.168.15.35/24.
Gigabitethernet 1/0: 192.168.164.80/24.

Juniper vMX

Em4: 192.168.15.25 /24.

Mikrotik

Ether 1: 192.168.15.20/24.
Ether 2: 192.168.15.54 /24.

Fortigate Firewall

Port 1: 192.168.15.40 /24.

We have initiated the configuration of the cactus monitoring system in the Centos-Linux server after allocating IP addresses.

At first, we installed cacti required packages and software in the cantos server by using some commands. Then we installed HTTPD (webserver), MariaDB (database Server) and Php server in centos.

After that, we configured SNMP (Simple Network Management Protocol) services in all devices and then checked the SNMP status of the devices in the Centos server. Simple Network Management Protocol is a protocol that is used for information collection and network device configuration. It typically operates on UDP port 161. The SNMP service is used extensively in network monitoring tools to display current device and interface status, as well as bandwidth and packet loss status. Before installing the cactus tool, we also installed RRDTool and NTP (Network Time Protocol) (Fig. 4).

After completing the configuration of the cacti server, we also configured the weather map and smokeping monitoring tools in the cento's server.

For the weather map, we installed a git clone from GitHub and then added a plugin in cacti. Following that, we added devices to the weather map editor and connected them with IP addresses and device interfaces to create our weather map tool (Fig. 5).

For Smokeping, firstly, other software packages, EPEL, development tools, and a Perl library that are specified as smokeping Dependencies in the Centos server must be installed. Additionally, we configured smokeping on Centos server by adding devices IPDNS IP and different types of presentation components such as median, maximum and minimum bandwidth, packet loss etc. (Fig. 6).

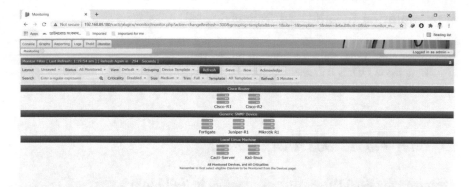

Fig. 4 Cacti tool

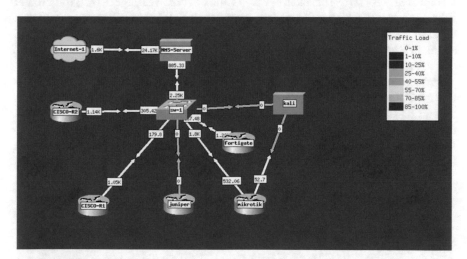

Fig. 5 Weather map tool

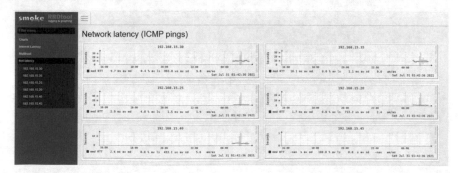

Fig. 6 SmokePing Tool

4   Simulation Results and Discussion

We applied total 2 attacks in our system model network.
1. Ping flood DDoS Attack.
2. TCP SYN Flood DDoS Attack.

4.1   Ping Flood DDoS Attack Output

Executed Ping flood DDoS Attack command in Kali Linux

In the Fig. 7, we created Ping flood, also known as ICMP flood which is a common
Denial of Service (DoS) attack in our Kali Linux Server. We targeted Mikrotik router
Ethernet 2 port IP which is directly connected with Kali Linux Ethernet 1 port. Here,
we used the "hping3 -1 –faster 192.168.15.54" command to create a ping flood DDoS
attack in Kali Linux. From this figure, we see that nearly 84% packet loss happens
due to that as it overloads the network with target data packets.

After executing the ping flood DDoS attack in Kali Linux, we can see from Fig. 8
that the Output of traffic load in the weather map. It clearly shows that traffic load was
around 10–25% in the kali Linux to Mikrotik route within a short time as the attack
created numerous amounts of ping requests. It clearly shows that other interfaces
look normal in this network while Mikrotik router Ethernet 2 port gets a huge traffic
load due to that ping flood DDoS attack.

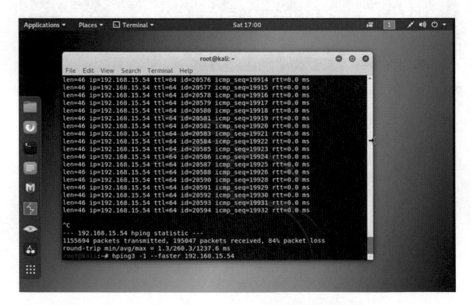

Fig. 7 Ping flood DDoS attack command in Kali Linux targeted to Mikrotik router

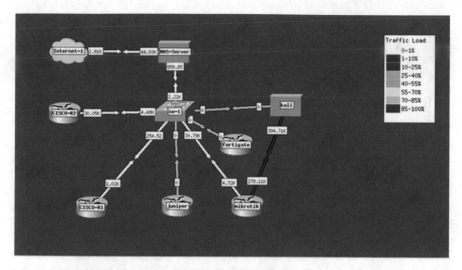

Fig. 8 Output of Ping flood DDoS Attack in Weathermap

Afterwards, login into the Mikrotik router from the winbox tools of our computer, from Fig. 9 we can see the output of the rate of traffic passes in the ether2 interface of the Mikrotik router. It shows how a simple dos attack command can create huge bandwidth flows in our network device. As a result, the device could be hung or the interface could be down anytime due to excessive traffic load which can create massive problems in any core network.

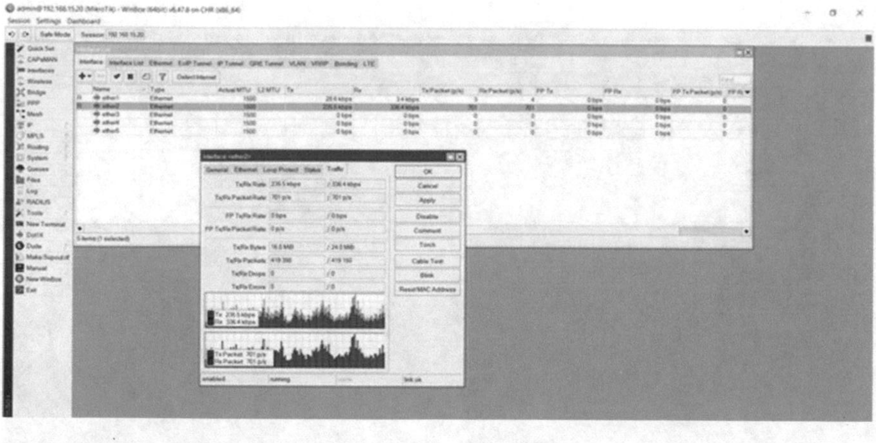

Fig. 9 Output of Ping flood DDoS Attack in Mikrotik Router

```
            :-# hping3 -S --flood -p 80 192.168.15.20
HPING 192.168.15.20 (eth0 192.168.15.20): S set, 40 headers + 0 data bytes
hping in flood mode, no replies will be shown
^C
--- 192.168.15.20 hping statistic ---
652130 packets transmitted, 0 packets received, 100% packet loss
round-trip min/avg/max = 0.0/0.0/0.0 ms
            :-#
```

Fig. 10 TCP SYN Flood DDoS attack in kali targeted to Mikrotik router

4.2 *TCP SYN Flood DDoS Attack Output:*

Execute TCP SYN Flood DDoS Attack command in Kali Linux

According to Fig. 10, we executed TCP SYN Flood DDoS Attacks in the network
to show how a DDoS attack can be more destructive for a network. We applied this
attack in our kali-Linux machine by using the command "hping3 –s –flood –p 80
192.168.15.20" which targeted the Mikrotik router ethernet1 port that is directly
connected to switch 1. Here we can see that almost 100% packet loss occurs due
but that sends TCP connection requests faster than the targeted machine can process
them, causing network saturation (Fig. 11).

After executing the TCP SYN Flood DDoS attack in Kali Linux, we can see the
Output of traffic load in the weather map. It clearly shows that the enormous traffic
load was around 10–25 within the short time. It clearly shows that other interfaces
look normal in this network while Mikrotik router Ethernet 1 port gets a huge traffic
load due to that attack.

After login into the Mikrotik router from the winbox tool of our computer, from
Fig. 12 we can see the output of the rate of traffic passes in the ether1 interface of

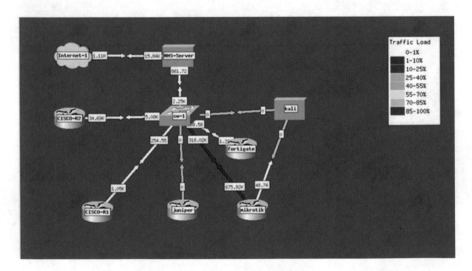

Fig. 11 Output of TCP SYN Flood DDoS attack in Weathermap

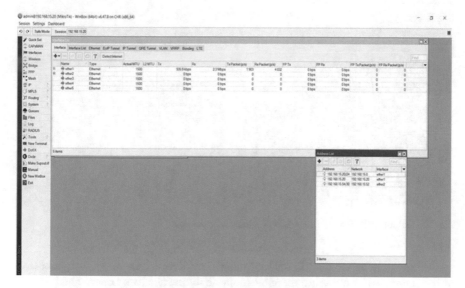

Fig. 12   Output of TCP SYN Flood DDoS Attack in Mikrotik Router

the Mikrotik router. It shows how a simple DDOS attack command can create huge bandwidth flows in our network device. It can be more destructive than dos attack and the amount of traffic received by the port is eight times higher than the ping flood DDoS attack. Eventually, as the server's connection overflow tables fill, service to legitimate clients will be denied, and the server may even malfunction or crash.

Cacti monitoring tools show the real-time instant bandwidth of any interface. After executing the DDoS attack on the Mikrotik device (Fig. 13), suddenly traffic flow increases and it shows how inbound traffic creates an immediate effect on the Mikrotik router ether1 interface.

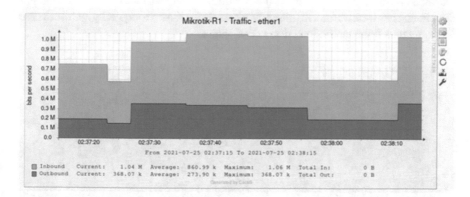

Fig. 13   Output of TCP SYN Flood DDoS Attack in Cacti

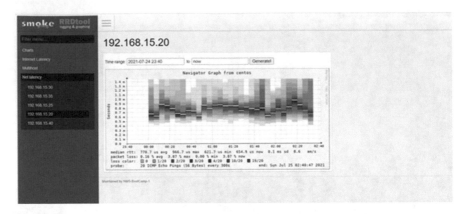

Fig. 14 Output of TCP SYN Flood DDoS Attack in smokeping

Before applying the DDoS attack on the Mikrotik router, there was no packet loss in the router. After applying the attack, suddenly packet loss occurs and that was seen in the smokeping graph (Fig. 14). Smokeping showed the immediate effect of DDoS attacks on the Mikrotik router and packet loss amount and 20% packet loss blue color shows the dangerous effect of DDoS attack on a network device.

We executed TCP SYN Flood DDoS Attacks in the network (Fig. 15) to show how a DDoS attack can be more destructive for a network. We applied this attack in our kali-Linux machine by using the command "hping3 –s –flood –p 80 192.168.15.30" which targeted the Cisco R1 router gigabit ethernet 0/0 interface that is directly connected to switch 1. Here we can see that almost 100% packet loss occurs due but that sends TCP connection requests faster than the targeted machine can process them, causing network saturation.

After executing the TCP SYN Flood DDoS Attack in Kali Linux, we can see the Output of traffic load in the weather map (Fig. 16). It clearly shows that the enormous traffic load was around 10–25 within the short time. It clearly shows that other interfaces look normal in this network while the Cisco router R1 gigabit ethernet 0/0 port gets a huge traffic load due to that attack.

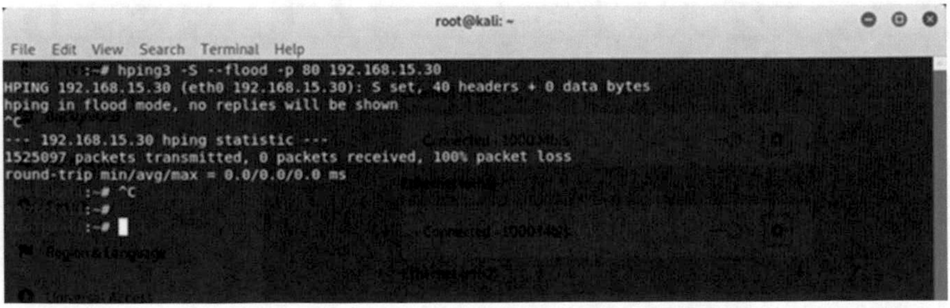

Fig. 15 TCP SYN Flood DDoS attack in kali targeted to Cisco R1 router

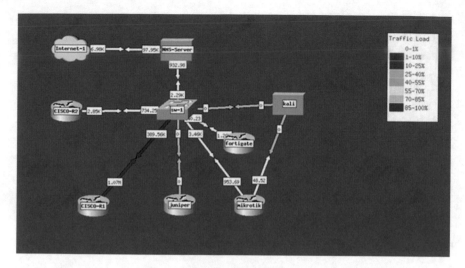

Fig. 16 Output of TCP SYN Flood DDoS Attack in Weathermap

Cacti monitoring tools show the real-time instant bandwidth of any interface. After executing the DDoS attack on the Cisco R1 device (Fig. 17), suddenly traffic flow increases and it shows how inbound traffic creates an immediate effect on the Cisco R1 router gigabit Ethernet 0/0 interface.

Before applying the DDOS attack on the R1 router, there was no packet loss in the router. After applying the attack, suddenly packet loss occurs and that was seen in the smokeping graph (Fig. 18). Smokeping showed the immediate effect of DDOS attacks on the Cisco R1 router. Packet loss amount and 20% packet loss blue color shows the dangerous effect of DDOS attack on a network device.

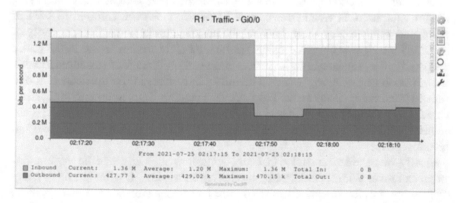

Fig. 17 Output of TCP SYN Flood DDoS Attack in Cacti

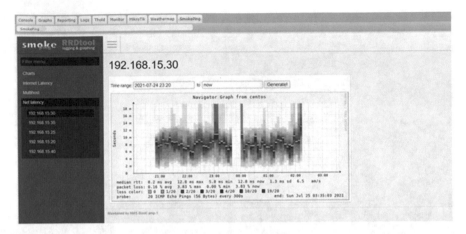

Fig. 18 Output of TCP SYN Flood DDoS Attack in smokeping

5 Conclusion

In our research, we have successfully presented the severity of DoS and DDoS attacks by applying Ping flood and SYN-flood attacks. A Linux server-based network monitoring system was built to run, incorporate and observe the effects of the DoS attacks DDoS attacks. Our findings show that a Ping flood attack leads to almost 84% of packet loss as the system was overloaded with data packets. When observed in Weathermap we can see that the traffic load was around 10–20%. Moreover, the winbox tool clearly shows how a simple DoS attack creates an overflow of bandwidth which results in the system crashing or disrupt. Executing TCP SYN flood attack portrays a more severe scenario, a 100% packet loss as TCP connection frequently sends faster connection requests that the victim machine cannot process, causing network saturation. In Weathermap, the SYN flood attack shows around 10–25% of traffic load within a very short time, but the interface looks normal. The severity of the DDoS attack was observed using Winbox, an eight-time increase in the traffic compared to the ping flood attack. Such overload of traffic can lead to malfunctioning or even crash of the target system. The immediate effect in the overload of traffic was observed using Cacti as well as using Smokeping. Implementing our system, which is a combination of AI-based and firewall network traffic monitoring system, we have monitored and detected the severe impact DoS and DDoS attack have and the risk it poses.

Finally, we have demonstrated that our conclusions are wider than the single application can give. With tools like Nagios, Nfsen, LibreNms, LogServer, this system architecture may be integrated in the future and many more deep threats will be demonstrated in these tools. Besides, an additional modelling approach could be included to demonstrate how an intrusion detection and prevention system can defend a network from threats and vulnerabilities. In future, we will work on a larger project involving both the monitoring of Network invasion and the identification output through various monitoring tools as well as network attack prevention via the next firewall generation.

References

1. A. Shenfield, D. Day, A. Ayesh, Intelligent intrusion detection systems using artificial neural networks. ICT Express **4**, 95–99 (2018). https://doi.org/10.1016/j.icte.2018.04.003
2. K. Valentín, M. Malý, Network firewall using artificial neural networks. Comput. Inf. **32**, 1312–13273 (2013)
3. A.P. Veiga, Applications of artificial intelligence to network security (2018). arXiv:180309992 [cs]
4. R. Radha, A survey on ai with cyber security. Int. J. Comput. Sci. Mob. Comput. (IJCSMC) **8**, 13–17 (2019)
5. N. Ben-Asher, C. Gonzalez, Effects of cyber security knowledge on attack detection. Comput. Hum. Behav. **48**, 51–61 (2015). https://doi.org/10.1016/j.chb.2015.01.039
6. E. Tyugu, Artificial intelligence in cyber defense, in *IEEE Xplore*. IEEE, Tallinn, Estonia (2011), pp. 1–11
7. H.M. Rajan, S Dharani, Artificial intelligence in cyber security – an investigation. Int. Res. J. Comput. Sci. (IRJCS) **4**, 28–30 (2017)
8. D. Patil, Artificial intelligence in cybersecurity. Int. J. Res. Comput. Appl. Robot. **4**(5), 1–5 (2016)
9. A. Panimalar, G. Pai, S. Khan, Artificial intelligence techniques for cybersecurity. Int. Res. J. Eng. Technol. (IRJET), **5**(3) (2018)
10. K.R. Bhatele, H. Shrivastava, N. Kumari, The role of artificial intelligence in cyber security, in *Countering Cyber Attacks and Preserving the Integrity and Availability of Critical Systems* (2019) , pp. 170–192. https://doi.org/10.4018/978-1-5225-8241-0.ch009
11. P. Vähäkainu, M. Lehto, Artificial intelligence in the cyber security environment, in *The Proceedings of the International Conference on Cyber Warfare and Security* (2019)
12. G.I.P. Duppa, N. Surantha, Evaluation of network security based on next generation intrusion prevention system. TELKOMNIKA (Telecommun. Comput. Electron. Control) **17**, 39 (2019). https://doi.org/10.12928/telkomnika.v17i1.9191
13. J. Jabez, B. Muthukumar, Intrusion detection system (IDS): anomaly detection using outlier detection approach. Proc. Comput. Sci. **48**, 338–346 (2015). https://doi.org/10.1016/j.procs.2015.04.191
14. M. Stampar, K. Fertalj, Artificial intelligence in network intrusion detection, in *IEEE Xplore* (2015), pp. 1318–1323. https://doi.org/10.1109/MIPRO.2015.7160479
15. K. Napanda, H. Shah, L. Kurup, Artificial intelligence techniques for network intrusion detection. Int. J. Eng. Res. **V4**(11) (2015)
16. Z. Ahmad, A. Shahid Khan, C. Wai Shiang, J. Abdullah, F. Ahmad, Network intrusion detection system: a systematic study of machine learning and deep learning approaches. Trans. Emer. Telecommun. (2020)
17. G. Kumar, K. Kumar, M. Sachdeva, The use of artificial intelligence-based techniques for intrusion detection: a review. Artif. Intell. Rev. **34**, 369–387 (2010). https://doi.org/10.1007/s10462-010-9179-5
18. M. Manninen, Using artificial intelligence in intrusion detection systems, in *CiteSeer*. http://citeseerx.ist.psu.edu/viewdoc/summary?doi=10.1.1.332.6432. Accessed 14 Oct 2021
19. R. Koch, Towards next-generation Intrusion Detection, Infona.pl, 2021. https://www.infona.pl/resource/bwmeta1.element.ieee-art-000005954707

20. C. Mazzariello, R. Bifulco, R. Canonico, Integrating a network IDS into an open source Cloud Computing environment, in *2010 Sixth International Conference on Information Assurance and Security* (2010). https://doi.org/10.1109/isias.2010.5604069
21. T. Matma, Anomaly based intrusion detection system: IJECI **3**, 7 (2019)
22. P.P. Mukkamala, S. Rajendran, A survey on the different firewall technologies. Int. J. Eng. Appl. Sci. Technol. **5**, 363–365 (2020). https://doi.org/10.33564/ijeast.2020.v05i01.059
23. E. Kolawole, W. Ali, O. Odejide, J. Fuller, Security issues, threats and possible solutions in cloud computing. Am. J. Inf. Sci. Comput. Eng. **5**(2), 38–46 (2019)
24. C. Modi, D. Patel, B. Borisaniya et al., A survey of intrusion detection techniques in Cloud. J. Netw. Comput. Appl. **36**, 42–57 (2013). https://doi.org/10.1016/j.jnca.2012.05.003
25. D. Savita, S. Kumar, A review of cloud computing environment and security challenges. Int. J. Adv. Res. Comput. Eng. Technol. (IJARCET) **7**(2), 174–182 (2018)
26. N.M.A. Al-Jaser, A survey on cloud computing security challenges and trust issues. Int. J. Comput. Sci. Inf. Secur. (IJCSIS) **18**(5), 7–12 (2020)
27. J. Kumar, Cloud computing security issues and its challenges: a comprehensive research. Int. J. Recent Technol. Eng. (IJRTE) **8**(1S4),10–14 (2019)
28. S. Shamshirband, M. Fahi, A. Chronopoulos, A. Montieri, F. Palumbo, A. Pescape, Computational intelligence intrusion detection techniques in mobile cloud computing environments: review, taxonomy, and open research issues. J. Inf. Secur. Appl. (2019)
29. T.-S. Chou, Security threats on cloud computing vulnerabilities. Int. J. Comput. Sci. Inf. Technol. **5**, 79–88 (2013). https://doi.org/10.5121/ijcsit.2013.5306
30. W.A. Varsha, S. Gupta, Study of security issues in cloud computing. Int. J. Comput. Sci. Mob. Comput. **4**(6), 230–234 (2015)
31. P. Chakraborty, M.D. Zahidur, S. Rahman, Building new generation firewall including artificial intelligence. Int. J. Comput. Appl. **178**, 1–7 (2019). https://doi.org/10.5120/ijca2019919416
32. H. Abie, An overview of firewall technologies (2000)
33. S. Dandamud, E. ProfT, Firewalls implementation in computer networks and their role in network security. J. Multidiscip. Eng. Sci. Technol. (JMEST) **2**, 408–411 (2015)
34. A. Tekerek, O. Bay, Design and implementation of an artificial intelligence-based web application firewall model. Neural Netw. World **29**(4), 189–206 (2019)
35. I. Kashefi, M. Kassiri, A. Shahidinejad, A survey on security issues in firewalls: a new approach for classifying firewall vulnerabilities, in *CiteSeer*. http://citeseerx.ist.psu.edu/viewdoc/summary?doi=10.1.1.415.4272. Accessed 14 Oct 2021
36. M. Imran, A. Alghamdi, B. Ahmad, Role of firewall technology in network security. Int. J. Innov. Adv. Comput. Sci. 3–6 (2015)
37. X. Liu, H. Lun, M. Fu, Y. Fan, L. Yi, W. Hu, Q. Zhuge, AI-based modeling and monitoring techniques for future intelligent elastic optical networks. Appl. Sci. **10**(1), 363 (2020)
38. Z. Mirza, Z. Farooqui, M. Jain, A. Karolia, Cloud-based solution for small and medium franchisees. Int. J. New Technol. Res. (IJNTR) **3**(3), 34–38 (2017)
39. E. Onyema, N. Eucheria, U. Nneka, R. Afriyie, O. Nwoye, Cloud security challenges: implication on education. Int. J. Comput. Sci. Mob. Comput. **9**(2), 56–73 (2020)
40. M.I. AL Ladan, Security issues in cloud-based businesses, in *9th International Conference on Computer Science, Engineering and Applications (CCSEA 2019)* (2019). https://doi.org/10.5121/csit.2019.90929
41. I. Gul, M. Hussain, Distributed cloud intrusion detection model. Int. J. Adv. Sci. Technol. **34**, 71–82 (2011)
42. S. Bandyopadhyay, R. Chowdhury, P. Banerjee, et al., A decision tree based intrusion detection system for identification of malicious web attacks (2020). https://doi.org/10.20944/preprints202007.0191.v1

43. M. Rani, Gagandeep, A review of intrusion detection system in cloud computing. SSRN Electron. J. (2019). https://doi.org/10.2139/ssrn.3355127
44. S. Devi, D.A.K. Sharma, Understanding of intrusion detection system for cloud computing with networking system. Int. J. Comput. Sci. Mob. Comput. **9**(3), 19–25 (2020)
45. S. Moorthy, M., & M, Rajeswari., Virtual host-based intrusion detection system for cloud. International Journal of Engineering and Technology (IJET) **5**(6), 5023–5029 (2013)

Chapter 7
Leveraging Artificial Intelligence Capabilities for Real-Time Monitoring of Cybersecurity Threats

Nickson M. Karie, Nor Masri Bin Sahri, Wencheng Yang, and Michael N. Johnstone

Abstract The number of cybersecurity incidents perpetrated by adversaries using modern complex and emerging technologies is growing exponentially. To curb this threat, advanced countermeasures need to be developed and implemented to mitigate cybersecurity incidents and detect the activities of adversaries. Developing advanced, innovative, and effective countermeasures is a challenge when faced with a daily tide of cybersecurity threats and concomitant vulnerabilities which, in most cases, have significant consequences to organisations. Artificial Intelligence (AI), which has specific, established use-cases in the cybersecurity domain offers promising solutions. This chapter examines how AI can enhance the real-time monitoring of cybersecurity threats in different environments. As a result, different cybersecurity threats, as well as commonly exploited cybersecurity vulnerabilities, are described and discussed in this chapter. Following this, some current real-time cybersecurity monitoring tools are discussed. Finally, the chapter highlights the role of AI in real-time monitoring focusing on how Explainable Artificial Intelligence (XAI) can be used to enhance real-time monitoring of cybersecurity threats which has become a crucial component of modern-day security implementations.

Keywords Cyber security threats · Real-time monitoring · Artificial intelligence

N. M. Karie (✉) · N. M. B. Sahri · W. Yang · M. N. Johnstone
Security Research Institute, School of Science, Edith Cowan University, Cyber Security
Cooperative Research Centre, Joondalup, WA, Australia
e-mail: n.karie@ecu.edu.au

N. M. B. Sahri
e-mail: n.sahri@ecu.edu.au

W. Yang
e-mail: w.yang@ecu.edu.au

M. N. Johnstone
e-mail: m.johnstone@ecu.edu.au

1 Introduction

The technological revolution of this century kickstarted a wave of smart technologies, the use of which has become the norm for individuals and organisations. This revolution is significant because while people want faster and more intelligent ways of solving problems, there is often an (unwarranted) expectation that security is not as important as functionality.

Smart technologies have influenced the way we communicate and interact with one another, possibly because of the proliferation of connected devices. The amount of information being shared on connected devices is phenomenal. However, hyper-connectivity and digitization have brought with them new and advanced cybersecurity threats. The world is experiencing an increasing volume of cybersecurity attacks–individuals, government agencies and organisations are exposed to these threats, which drives the need to invent new and advanced cybersecurity countermeasures to deal with existing and future cybersecurity threats and vulnerabilities.

Knowing that modern-day life is almost entirely dependent on technology, the realization of cybersecurity threats (henceforth referred to as "cyber threats" or simply "threats") may result in problems worse than simple disruption of services. The Information Security Forum [1] noted in its 2019 study of the cybersecurity landscape, the increased potential for:

- disruption from an over-reliance on fragile connectivity.
- distortion as trust in the integrity of information is lost; and
- deterioration, when controls are eroded by regulations and technology.

Therefore, for organisations to be able to defend their systems (computers, servers, mobile devices, electronic systems, networks) and data against existing and future malicious attacks, security systems need to evolve as well. This also implies that, as attackers are turning to more sophisticated attack methods, organisations of all sizes must as well embrace equally sophisticated cybersecurity countermeasures capable of monitoring, detecting, and responding to real-time cyber threats. This is because despite some threats being very hard to detect, there will often be symptomatic evidence for an organisation to notice after a cyber-attack. Unfortunately, the dwell time (or time difference between a cyber-attack and when an organisation notices the incursion into its network) might be over a year in some cases.

Given the dynamic nature of modern cyber threats, individuals, as well as organisations, will be disadvantaged if they are unable to determine their security posture in real-time [2]. For this reason, a combination of multiple countermeasures, ranging from advanced AI-based monitoring tools with explainable artificial intelligence capabilities to a workforce that is trained and aware of modern threat techniques, and vulnerabilities must be deployed for successful ongoing cybersecurity defence. This chapter thus aims to shed light on how AI (including explainable artificial intelligence) can be used to enhance the real-time monitoring of cybersecurity threats which has become a crucial component of modern-day security implementations [2].

In the remainder of this chapter, Sect. 2 presents a brief explanation of concepts background while Sect. 3 discusses common cybersecurity threats. Section 4 explains in brief sampled common cybersecurity vulnerabilities followed by a discussion of real-time monitoring tools in Sect. 5. Section 6 presents the role of AI in the real-time monitoring of cybersecurity threats focusing on XAI. Finally, Sect. 7 considers future research work.

2 Concepts and Background

This section covers Explainable AI (XAI) which focuses on the understanding and interpretation of the behaviour of AI systems and highlights the difference between AI and XAI. Cyber Security Threats is also discussed in his section to help spark new discussion on the development of advanced, innovative, and effective countermeasures.

2.1 Explainable AI (XAI)

Explainable Artificial Intelligence (XAI) according to [3] is a field that is concerned with the development of new methods that explain and interpret machine learning, deep learning, and neural networks models to help in the understanding and interpretation of the behaviour of Artificial Intelligence (AI) systems. XAI consists of a set of processes and methods that help human users to understand and trust the results generated by machine learning algorithms. XAI has in recent years been adopted for use in different application areas including healthcare, cybersecurity, and self-driving cars. Many reasons exist for the adoption of XAI, however, the primary driver is the need for trustworthy, fair, robust, and high performing models for real-world applications [3, 4]. As noted by [3] systems whose decisions cannot be understood are difficult to trust.

XAI is thus important in helping organisations build trust and confidence when developing AI models in their products and services, thus XAI can help promote user trust in AI. Further, the concept of explainability can assist developers to ensure that systems perform as expected and meet regulatory and legal standards. This last point is crucial to user trust as it provides an opportunity for parties affected by a decision generated using AI models to challenge or change that outcome [5]. With XAI, it is also possible for an organisation or individual to troubleshoot and improve model performance and at the same time help other stakeholders understand the behaviours of AI models. In the next sub-section, the difference between AI and XAI is highlighted.

2.2 Difference Between AI and XAI

Artificial Intelligence according to [6] is a field that attempts to build enhanced intelligence into computer systems. Artificial intelligence thus simulates human intelligence processes using machines, especially computer systems. The primary objective of AI is to enable computers and machines to perform intellectual tasks such as problem-solving, decision making, perception, and understanding human communication. Some of the more specific applications of AI include expert systems, natural language processing, speech recognition and machine vision.

In contrast, Explainable Artificial Intelligence is primarily concerned with the development of new methods that explain and interpret machine learning models to aid in the understanding and interpretation of the behaviour of AI systems. This, therefore, implies that XAI focuses on describing existing AI models, their expected impact, and potential biases. XAI assists organisations in troubleshooting and improving existing AI model performance but at the same time helps stakeholders understand the behaviours of the AI models. The next sub-section will cover cyber security threats.

2.3 Cyber Security Threats (CST)

Taylor [7] defines Cyber Security Threats (CST) as malicious acts by adversaries or processes that seek to damage data, steal data, or disrupt digital systems in general. The realisation of any cybersecurity threat may include but is not limited to, theft of valuables, electrical blackouts, failure of military equipment, breaches of national security, theft of sensitive medical data records, disruption of phone or mobile networks, interference with computer networks or making data unavailable.

With increased global connectivity and advances in Artificial Intelligence (AI), computers and other digital systems are becoming smarter and better connected [8]. Unfortunately, these developments are making countering CST difficult as adversaries are also leveraging the capabilities of the same technologies to make CST heterogeneous and hard to detect. One notable example is "zero-day" threats that can surprise cybersecurity defence systems because they are not known in advance (by definition) and therefore signature-based methods cannot match them as they do not carry detectable digital signatures. However, of more concern is a host of new and evolving threats such as the "Advanced Persistent Threats" (APT). According to [9] "Advanced Persistent Threat refers to a well-organized, malicious group of people who launch stealthy attacks against computer systems of specific targets, such as governments, companies or military". Adversaries can use APTs to gain ongoing access to sensitive data and may remain undetectable by traditional security measures for a long time.

To counter modern CST, we elaborate on leveraging AI capabilities for real-time monitoring of cybersecurity threats. This chapter was motivated by the fact that adversaries are often the first to embrace new technologies to help them discover new ways to annoy, damage or steal data, disrupt digital systems, or even harm individuals. In doing this, adversaries spend time to extensively profile targets to understand their weaknesses and vulnerabilities before exploiting them. For this reason, government agencies and organisations will need to embrace new, advanced, and dynamic ways to measure their security posture as well as constantly monitoring for new CST, security misconfigurations among many other vulnerabilities.

In this chapter, we agree with the views of [5] that AI has very distinct use-cases in the cybersecurity domain, and even if it's not a perfect solution, it can help enhance the fight against CST in different environments [10]. AI can also be used to develop new and advanced ways of monitoring CST to help security professionals, CISOs and other stakeholders make intelligent decisions as well as counter the rate of intrusions into government and private computer networks. The next section explains some of the common cybersecurity threats sampled to support the work presented in this chapter.

3 Common Cybersecurity Threats

As the cybersecurity threat landscape continues to evolve, new CST emerges almost daily [11]. In this section, some common cybersecurity threats are explained to facilitate this study. Ransomware is discussed first because it is currently the most prevalent threat. This is followed by other common cybersecurity threats including Data Leakage and Data Breach, Phishing, Social Engineering, Hacking, Insider threat, Denial of Service (DoS), Crypto-jacking, Cyber-Physical Attacks, Endpoint Attacks, Mobile Malware, and IoT Attacks.

3.1 Ransomware

Ransomware is a form of malware that targets both human and technical weaknesses to make critical data and/or systems inaccessible [12]. It can be delivered through various attack vectors including phishing emails and remote desktop protocol, which allows computers to connect across a network, then spreads throughout the network by installing malicious software. In most cases, adversaries performing ransomware attacks usually take control of a computer, lock its data, and demand a ransom from the victim promising to restore access to the data upon payment (usually by some form of untraceable cryptocurrency). When a compromised system is infected, the ransomware program shows a message requiring payment for functionality restoration. Once the payment is transferred, the decryption key will be provided, allowing the system to be restored to normal. However, there is no guarantee that even after the payment is done that everything will be restored to normal as it was before the attack. Note that, ransomware variants as described by [13] take many forms and their names keep changing from time to time, but some of the popular or common ransomware variants include [13]:

1. **Crypto Ransomware or Encryptors**: A popular variant that encrypts files and data within a system, making them inaccessible without a decryption key.
2. **Lockers**: This type of ransomware completely locks victims out of a system making files and applications inaccessible. A lock screen displays the ransom demand, possibly with a countdown clock to increase urgency and drive victims to act.
3. **Scareware**: Fake software that claims to have detected a virus or other issue on a computer and directs the victim to pay to resolve the problem. Some types of scareware lock the computer, while others simply flood the screen with pop-up alerts without damaging files.
4. **Doxware or leakware**: This type of ransomware threatens to distribute sensitive personal or company information online, and many victims panic and pay the ransom to prevent private data from falling into the wrong hands or entering the public domain. Another variation is **police-themed ransomware**, which claims to be law enforcement and warns that illegal online activity has been detected, but jail time can be avoided by paying a fine.
5. **RaaS (Ransomware as a Service)**: This refers to malware hosted anonymously by a professional adversary that handles all aspects of the attack, from distributing ransomware to collecting payments and restoring access, in return for a percentage of the profit as a service fee.

3.2 Data Leakage and Data Breach

Skrop [14] defines data leakage as an uncontrolled or unauthorised transmission of classified information to the outside world. Data leakage can also mean the unauthorised transmission (electronically or physically) of data from within an organisation to an external destination or recipient [15]. Many software solutions have been developed to handle data protection. However, existing data leakage detection systems cannot provide absolute protection hence the need to develop new intelligent data leakage detection systems [16]. Data breach on the other hand is a security incident in which information is accessed without authorization. This occurs when government or organisation confidential, private, or other sensitive information is accessed without authorisation or in some cases data is lost. In a recent data breach, LinkedIn saw data associated with more than 700 million of its users posted on a dark web forum in June 2021, impacting more than 90% of its user base.[1] In this case, hackers used data scraping techniques to exploit the LinkedIn API before dumping the customers' data. While LinkedIn argued that no sensitive or personal data were exposed and was therefore simply a violation of its terms of service, the incident is

---

[1] https://www.csoonline.com/article/2130877/the-biggest-data-breaches-of-the-21st-century.html.

still considered a data breach. Knowing how data leakage and data breach occur in practice can help protect organisational data.

Some countries have also well-established data protection laws. In Australia, for example, the Office of the Australian Information Commissioner (OAIC) is an independent national regulator for privacy and freedom of information.[2] The OAIC promotes and upholds the rights of individuals to access government-held information and has their personal information protected. Other areas under the regulation of OAIC include privacy rights, credit reporting, Australian Privacy Principles guidelines, Notifiable data breaches and the Australian Government Agencies Privacy Code. The OAIC also looks after Freedom of information (FOI) rights, Access information under FOI, FOI Guidelines as well as the FOI Act.

In Europe, there is the General Data Protection Regulation (GDPR)[3] which covers data protection and privacy in the European Union and the European Economic Area.

3.3 *Phishing*

Phishing is one of the oldest ways of stealing information, especially personal details such as user credentials and online banking information [17]. According to [18] phishing is a major threat to all Internet users and is difficult to trace or defend against since it deliberately does not present itself as obviously malicious. It has become a growing threat in cyberspace driven mainly by variations such as spear-phishing and whaling that are more convincing and thus harder to detect. Spear phishing for example can be understood as a scam driven by either email or electronic communications that target specific individuals, organisations, or businesses. Although often intended to steal data for malicious purposes, spear phishing can also be used by adversaries to install malware on a targeted user's computer. With spear phishing, the attackers usually disguise themselves as trustworthy friends or entities to the victim to acquire sensitive information, typically through email or other online messaging.[4] The use of email or other online messaging systems for spear phishing is considered the most successful form of acquiring confidential information on the internet, accounting for 91% of attacks.[5] Whaling, also known as CEO fraud, is a variation of phishing that uses similar methods to trick a target into performing specific actions, such as revealing sensitive data or transferring money.[6] The cyber-criminals usually masquerade as senior players at an organisation and directly target senior or other important individuals at the organisation, to steal money or sensitive information or gain access to computer systems for criminal purposes.

---

[2] https://www.oaic.gov.au/.

[3] https://gdpr-info.eu/.

[4] https://digitalguardian.com/blog/what-is-spear-phishing-defining-and-differentiating-spear-phishing-and-phishing.

[5] https://www.firmex.com/resources/infographics/spear-phishing-whos-getting-caught/.

[6] https://www.kaspersky.com/resource-center/definitions/what-is-a-whaling-attack.

Phishing continues to be a popular technique of exploitation used by hackers to compromise major industry networks [19]. Publicly shared industry information also can be used by hackers to design their phishing techniques and use advanced crafted malware to avoid detection. Research done by IBM revealed that in 2021, the average cost of a data breach is now at USD 4.24 M.[7] The Anti-Phishing Working Group (APWG) on the other hand found that in the 1st Quarter of 2021, financial institutions, social media, webmail, payment systems, eCommerce and retails, logistics and shipping including cryptocurrency remain among the most targeted sectors while email fraud, remains one of the most frequent ways in which cybercriminals obtain unauthorized access to their victims' information [20]. Other areas targeted by phishing attacks include but are not limited to government agencies, the military, and the healthcare industry.

3.4 Social Engineering

Social engineering is the art of manipulating victims so that they provide confidential or sensitive information to an adversary who is otherwise not authorised to hold the information [21]. Social engineering often involves the acquisition of either computer systems data or users' credentials through techniques that include in-depth non-technical means. It relies heavily on human interaction and involves manipulating people into breaking normal security procedures and best practices to gain access to systems, networks, or physical locations, or for financial gain [21]. Social engineering techniques are improving with special and carefully crafted attempts to their victim using individual personal and social media networks. The more sophisticated the techniques being used in the social engineering process, the more they decrease the ability of the user to be able to distinguish between legitimate and malicious requests. Some of the notable examples of social engineering include the high-profile Twitter users' accounts compromised after the Vishing scam and the Microsoft 365 phishing scam which was used to steal user credentials.[8] A brief description of each of these realized social engineering examples is given below.

3.4.1 High-Profile Twitter Users' Accounts Compromised[8]

In July 2020, Twitter lost control of some users' accounts. Among the accounts hijacked by the hackers included those of some high-profile users such as Barack Obama, Jeff Bezos, Elon Musk, Joe Biden, and Kanye West. The hackers downloaded users' Twitter data, messages, and tweeted using the hijacked accounts requesting donations to a Bitcoin wallet from followers in exchange for a doubling in value. The hackers managed to obtain $110,000 in Bitcoin after hundreds of transactions

---

[7] https://www.ibm.com/au-en/security/data-breach.

[8] https://www.ibm.com/au-en/security/data-breach.

were completed within minutes before Twitter discovered and remove the tweets. This incident was described by Twitter as a "phone spear phishing" (vishing) attack. Twitter employees were tricked into revealing account credentials that allowed access to the compromised accounts. The FBI launched an investigation into Twitter's security procedures following the hack that caused a fall in Twitter's share price by 7% in pre-market trading the following day.

3.4.2   Microsoft 365 Phishing Scam[8]

Unlike the Twitter scam, in this Microsoft incident, security researchers discovered a Business Email Compromise (BEC) scam that tricked recipients into installing malicious code on their devices. The hackers managed to send a blank email to their targets with a subject line indicating a "price revision." The email contained an attachment that appeared to be an Excel spreadsheet file (.xlsx). The attachment, however, was not a spreadsheet but a.html file in disguise. Upon opening the (disguised).html file, the target was directed to a website containing malicious code which triggered a pop-up notification, telling the user s/he has been logged out of Microsoft 365 and inviting them to re-enter their login credentials. On entering the login credentials, the fraudulent web form captured and sent the user's credentials to the hackers behind the scam. This type of phishing relied on human error combined with weak defences and has been used even during the COVID-19 pandemic period to scam many other users working from home.

3.5   *Hacking*

Hacking is the unapproved utilisation of computer and system assets or the act of changing computer equipment and programming to achieve an objective outside of the maker's unique reason [22]. This implies that an adversary may alter system or security features to accomplish a goal that differs from the original purpose of the system thus breaching one's privacy and accessing confidential information. With hacking, the weaknesses of a system or loopholes in a network are identified and private details accessed [23]. The adversary then gains access to personal information by using exploits to break into the network, computer, or mobile device. One common example of hacking is cookie theft.[9] Most websites make use of cookies to allow for a personalized web browsing experience. Web browsers on the other hand also use cookies to store user passwords, bookmarks and browsing history for faster surfing. If the websites are not encrypted or do not use HTTPS, adversaries can intercept user data and hijack a browsing session. This gives them access to the cookies, as well as login details which can be used to commit other malicious activities.

---

[9] https://softwarelab.org/what-is-hacking/.

3.6 Insider Threat

An insider threat is a malicious threat to an organisation that comes from people within an organisation. The insider will intentionally or unintentionally misuse his or her access rights to negatively affect the confidentiality, integrity, or availability of the organisation's critical information or systems. It can also be one or more individuals with the access and/or inside knowledge of a company, organisation, or enterprise that would allow them to exploit the vulnerabilities of that entity's security, systems, services, products, or facilities with the intent to cause harm [24]. Insiders usually can identify and understand businesses vulnerabilities and they know how and when the business can be exploited. One example was Postbank,[10] South Africa's Post Office Bank, which was the victim of a major insider security breach. In this incident, multiple employees copied the master encryption key giving them access to Postbank customer accounts as well as the bank's internal systems. With this access, they were able to view account balances, reset bank cards, exfiltrate personal information, and steal money. The breach affected 12 million cards and eight to ten million people relying on Postbank for government assistance funds. It took Postbank a year to discover the full extent of the fraud. The hack amounted to around 25,000 fraudulent transactions and USD 3.35 million in damages. More examples of Insider Threat-caused breaches can be found at [25].

3.7 Denial of Service (DoS) and Distributed Denial of Service (DDoS) Attack

A DoS attack is a specially crafted attack designed to make the victim's computer or server inaccessible, usually by flooding the victim's system with access requests [26]. By doing this, the system's services will not be available to legitimate users (the victims). A modern DoS attack often relies on a botnet to attack a victim in a distributed way, therefore a Distributed DoS (DDoS). This may include, flooding the target system with traffic, or sending it information that triggers a crash. [26]. The primary difference is that, instead of a system being attacked from one location, the target is attacked from many different locations at once often distributed globally in what is referred to as a botnet [27]. One famous example is the Mirai botnet which took advantage of insecure IoT devices such as IP cameras and home routers to scan the Internet for open Telnet ports. It then turned the compromised devices into remotely controlled bots that could be used as part of a botnet in large-scale attacks. Adversaries could then try to log in using default passwords making it possible for the hackers to amass a botnet army.[11]

---

[10] https://www.proofpoint.com/us/blog/insider-threat-management/5-examples-insider-threat-caused-breaches-illustrate-scope-problem.

[11] https://www.csoonline.com/article/3258748/the-mirai-botnet-explained-how-teen-scammers-and-cctv-cameras-almost-brought-down-the-internet.html.

3.8 Crypto-Jacking

"Crypto-jacking is an emerging online threat or malware that hides on a computer or mobile device and uses the machine's resources to "mine" cryptocurrencies" [28]. Crypto-jacking can hijack web browsers, as well as compromise all kinds of devices, from desktops and laptops to smartphones and even network servers. Adversaries use crypto-jacking to steal computing resources from victims' devices to compete against sophisticated crypto mining operations without the costly overhead. In the end, it slows down computer processes, increases electricity bills as well as shortens the lifespan of devices [28]. Adversaries use malicious links in an email that loads crypto mining code on the computer, or by infecting a website or online advert with JavaScript code that auto-executes once loaded in the victim's browser.

3.9 Cyber-Physical Attacks

According to Scheuermann [29], a cyber-physical attack on critical infrastructure occurs when an adversary gains access to a computer system that operates equipment in a manufacturing plant, oil pipeline, a refinery, an electric generating plant, or other similar infrastructure and can control the operations of that equipment to damage those assets or other property [29]. Cyber-physical attacks are a risk not only for the owners and operators of those assets, but also for their suppliers, customers, businesses, and persons in the vicinity of the attacked asset, and any person or entity that may be adversely affected by it [29]. With cyber-physical attacks, an adversary could disable cameras, turn off a building's lights, make a car veer off the road, or a drone land in enemies' hands. Defences against this type of attack are becoming more challenging, due to the convergence of IT networks with Operational Technology networks, as noted by [30]. In recent times one of the examples of a cyber-physical attack happened the city of Kyiv, Ukraine. The incident caused a power outage at the Pivnichna remote power transmission facility. The grid hack was one of the thousands of large and small cyber intrusions into Ukrainian networks, suggesting that cyber-physical attacks are becoming an established part of the toolkit for state-sponsored actors.[12]

3.9.1 Endpoint Attacks

Endpoints can be defined as points of access to an organisation's network that create points of entry that can be exploited by adversaries. An endpoint therefore can be a local or remote computing device that communicates back and forth with a network to which it is connected [31]. This may include but is not limited to Point-of-sale, Industrial control or fixed-function devices, Desktops, Laptops, Smartphones, Tablets,

---

[12] https://securityledger.com/2017/01/six-cyber-physical-attacks-the-world-could-live-without/.

Servers, Workstations among other devices. Endpoints are increasingly becoming susceptible to cyberattacks thus making it easier for adversaries to bypass security measures. Endpoint security however is intended to protect these devices from malicious internal and external threats. More details of some endpoint threats can be found at [32].

3.9.2 Mobile Malware

Mobile malware is malicious software that specifically targets the operating systems of smartphones. This may include but is not limited to Spyware and Malware, Drive-by Downloads, Viruses and Trojans, Mobile Phishing and Browser Exploits. The goal of Mobile malware is to gain access to private data found on different phone devices. Some of the common examples of Mobile malware include:

3.9.3 DroidKungFu

DroidKungFu is mobile malware that can evade antimalware software detection and install a backdoor in the Android OS that allows adversaries to gain full control of the compromised mobile device.

3.9.4 Plankton

Plankton is mobile malware that can hide in existing apps. it can steal user browser data and could access a remote server to add more malicious files to the compromised device.

3.9.5 DroidDream

DroidDream is mobile malware that affects Android phones and can send user information to a remote server.

3.9.6 HongTouTou

HongTouTou is a worm that affects Android phones and was designed to generate search engine counts for the people who created it.

3.9.7 Geinimi

Geinimi spreads to devices using third-party apps found on the Android marketplace. This malware then sends out a user's location and app list to a remote server and can download apps as well.

3.9.8 Zitmo

Zitmo was initially designed to target Blackberry and Symbian devices to hack users' bank accounts but can target other phones as well.

3.9.9 IoT Attacks

Internet of Things (IoT) devices are a fast-growing sector of the Internet. However, as the number of connected devices proliferates, the increase creates additional points of access for an adversary. Nowhere is this more apparent than in the new domain space of smart cities. Baig et al. [33] state that smart cities are "a complex area, comprised of diverse and interconnected components constantly exchanging data and facilitating improved living for a nation's population. A typical smart city consists of four key components, namely, Smart Grids, Building Automation Systems (BAS), Unmanned Aerial Vehicles (UAVs), Smart Vehicles; with enabling Internet of Things (IoT) sensors and (often) a Cloud platform". It can be very difficult and challenging to immediately detect and mitigate each security threat posed by various poor security settings on the IoT devices deployed. This is because many of the IoT devices are manufactured without taking their security implications into account [34]. As noted by [35], Mirai was a particularly successful example of an IoT-based Botnet. More access means more opportunities for unwanted entry points. This is true of any smart building, but it is especially worrisome in the home, where everyone ought to feel safest [36]. The proliferation and increasing ubiquity of IoT devices have sparked a malicious attackers' renaissance, as most IoT devices require Web access or mobile apps for manual control.

Many other common cybersecurity threats exist that may not have been covered or beyond the scope of this chapter. The reader is encouraged to explore more literature for other existing cybersecurity threats including Pandemic-related and cloud-based cybersecurity threats, The next section presents some of the commonly known cybersecurity vulnerabilities.

4 Common Cyber Security Vulnerabilities

Vulnerability is a term used to refer to the weaknesses, unintended flaws in an organisation or in a set of procedures, software code or the computer system itself that leaves information security exposed in such a way that a threat actor, such as a hacker, nation-state, disgruntled employee, or other attackers, can exploit to adversely affect data security. Cyber Security Vulnerabilities (CSV) on the other hand, specifically focuses on weaknesses or unintended flaws related to information technology systems used by organisations, computing software as well as hardware [37]. Whichever the case, vulnerabilities leave an organisation, and its systems open to potential exploitation in the form of unauthorized access or malicious behaviour such as viruses, worms, trojan horses among other forms of malware. Some of the notable and common cyber-security vulnerabilities are briefly explained below. This is because vulnerabilities can be found in a variety of areas in an organisation or a system, including system hardware or software, weaknesses in policies and procedures used in the systems and weaknesses of the system users themselves [38].

4.1 Use of Broken Algorithms

The use of broken cryptographic algorithms can lead to the disclosure of sensitive information in an organisation. This is because a determined adversary may be able to break the algorithm and compromise any data that have been protected. This vulnerability can lead to compromised Confidentiality, Integrity, Availability as well as Accountability of any information owned by an organisation [39, 40].

4.2 Security Misconfiguration

Hardening an organisation's security requires defining and deploying a secure configuration for organisational applications such as web servers and database servers. However, in doing so, security misconfiguration vulnerabilities can occur if a component is susceptible to attack due to an insecure configuration option. This may be due to an insecure default configuration, poorly documented default configuration, or poorly documented side-effects of optional configuration. Examples include failing to set a useful security header on a web server or forgetting to disable default platform functionality that could grant administrative access to an adversary [41]. Some of the common security misconfigurations that can be exploited include unpatched systems, using default account credentials, unprotected files and directories, unused web pages, poorly configured network devices[13] among others.

---

[13] https://resources.infosecinstitute.com/topic/guide-preventing-common-security-misconfigura tions/.

4.3 Missing Authorization

Missing authorization is considered very dangerous according to [42] and occurs when a software or an application does not perform an authorization check when an individual attempts to access a resource or perform an action. An example would be allowing anyone to run arbitrary SQL queries on a given database attached to a victim machine and the database returning the result of the query without prompting for any authorisation. Missing authorisation can lead to the root cause of URL-guessing attacks [43], information exposure, DoS, arbitrary code execution among other attacks.

4.4 Missing Authentication for Critical Function (MACF)

MACF occurs when software does not perform any authentication, to any individual or a process, for an activity that requires a user identifier such as a username and a password or consumes a significant amount of computing resources such as memory or CPU [44].

4.5 Failure to Restrict URL Access

This occurs when an application fails to protect its web documents sufficiently and can lead to forced browsing which can be used to gather sensitive data through a web browser by requesting specific web pages or files. Failure to restrict URL access according to [45] may also enable attackers to forge URLs to access hidden pages or admin pages. This technique enables an adversary to bypass website security by accessing files directly instead of following designated links. This can happen mostly when a webpage does not have the correct access control policy in place hence unauthorized users can view content that they should not have the ability to view.[14] Failure to Restrict URL Access can also happen when all the privileges to access a webpage are checked on the client-side but not on the server-side. Adversaries can use personal proxies to bypass client-side privileges and access functionality not intended for them to access.

4.6 Insecure Cryptographic Storage

An Insecure Cryptographic Storage vulnerability occurs when sensitive organisational data (stored in files, databases, or other locations) is not secured [46].

---

[14] http://bretthard.in/post/restricting-url-access.

In most cases, some applications that process sensitive data and information are responsible for protecting the same data. However, when a failure of an application to protect data in storage using a recommended industry-standard encryption algorithm occurs then it leads to an Insecure cryptographic storage vulnerability which can be exploited by adversaries. For example, if a database contains a user's table with the password column saved using an MD5 hash of a password like "5f4dcc3b5aa765d61d8327deb882cf99" is exposed, a simple search of the hash value can reveal general and common passwords.

4.7 Insecure Direct Object References (IDOR)

Like SQL injection, this vulnerability occurs when an application provides direct access to objects such as database entries belonging to other users, files in a system based on user-supplied input. In this case, adversaries can bypass authorization and access resources in the system directly by modifying the value of a parameter used to directly point to an object. This happens mostly when an application takes user-supplied input and uses it to retrieve an object without performing enough authorization checks [47].

4.8 Unrestricted Upload of Dangerous File Types

This occurs when an application does not validate or improperly validates file types before uploading to any system. The application of the software allows adversaries to upload or transfer files of dangerous types that could be automatically executed within the application's environment [48, 49].

Other known cybersecurity vulnerabilities that can expose an organisation to cybersecurity threats include: Reliance on Untrusted Inputs in a Security Decision, Path traversal, Bugs, Weak Passwords, Software that is already infected with a virus, Broken authentication and session management, Download of codes without integrity checks, URL redirection to untrusted sites, Insufficient transport layer protection, Unvalidated redirects and forwards, Missing data encryption, OS command injection, SQL injection and Buffer overflow. The next section briefly explains selected cybersecurity real-time monitoring tools.

5 Real-Time Monitoring (RTM) Tools

Traditional monitoring systems are built to meet certain pre-defined compliance requirements and detect known attacks by using pre-defined rules. As good as traditional monitoring systems are, their static nature means they are ineffective at detecting some of the new and advanced CST. Moreover, modern cybersecurity attacks are sophisticated and often hidden, with the ability to bypass most of the pre-defined compliance requirements as well as pre-defined rules. For this reason, modern security monitoring tools need to go beyond normal traditional compliance monitoring using pre-defined rules, into a new dimension that offers real-time intelligent monitoring and detection of non-compliance using advanced tools. It is for this reason that the authors believe that AI can be used to enhance the real-time monitoring of cybersecurity threats in different organisations. The following sub-sections briefly explain selected well-known cybersecurity monitoring tools and their capabilities. In the context of this study, we divide the monitoring tools into two categories:

- Threat Maps Based Monitoring Tools
- Non-Threat Maps Based Monitoring Tools

5.1 Threat Maps Based Monitoring Tools

A cyber threat map, also known as a cyber-attack map, is a real-time map of the computer security attacks that are occurring at any given time. In more practical applications RTM needs to happen at the router level (or close to it) or can be part of an AI-enabled IDS which then enables real-time response. Some of the common threat map-based monitoring tools are discussed below.

- **ThreatCloud - Live Cyber Threat Map**[15]: Developed by Check Point Software Technologies this platform shows live statistics of different cybersecurity attacks including some of those discussed in Sect. 3, display the total amount of attack data, including new attacks, the source of the attacks and their destinations. This tool also can show the threats statistics for the past week or month, infection rate and top targeted countries and industries.
- **Cyberthreat Real-Time Map**[16]: Cyberthreat Real-Time map by Kaspersky can among other things scan, emails and perform intrusion detection and present a list of information captured by security vulnerabilities. Once the map is loaded, it detects the current location and shows the statistics for that country, including historical top local infections. Different features are present which display Botnet Activity Detection, Vulnerability Scan, Intrusion Detection Scan, Web Anti-Virus, Mail Anti-Virus, On-Demand Scan as well as On-Access Scan.
- **Darktrace Antigena**[17]: Darktrace Antigena is a tool that utilises machine learning to detect breaches in a network, Email, Cloud and SaaS. Darktrace uses an unsupervised machine learning algorithm that does not require previously identified and

---

[15] https://threatmap.checkpoint.com/.

[16] https://cybermap.kaspersky.com/.

[17] https://www.darktrace.com/en/darktrace-antigena/.

tagged data. The system detects breaches by actively combining similarity clustering techniques and anomaly monitoring. Antigena can also list similar connection patterns where a couple of devices are grouped and a similar pattern between the devices is shown. Darktrace further claims that its solution can detect and automatically respond to a cyber threat within a matter of seconds. The detection system uses physical and digital sensors on a client's network. The data captured from the sensors is channelled to over 60 different unsupervised AI learning algorithms that compete between themselves to determine any anomalous behaviour. Various statistical methods are used to process the input to differentiate which of the 60 to listen or to ignore. The system quarantines the detected breach until administrators decide a course of action.

- **Digital Attack Map**[18]: A collaboration between Google and Arbor Networks, this tool has live data visualization dedicated to tracking all incidents related to DDoS attacks around the world. The data gathered by this system come from a worldwide analysis of over 300 ISPs with over 130 Terabytes per second of live traffic.
- **FireEye Cyber Threat Map**[19]: FireEye Cyber ThreatMap like other threat maps can also show the origin, destination, total number of attacks and other statistics for the previous 30 days, such as top attacker countries and topmost attacked industries.

Other common monitoring tools that are also present in the industry include Bitdefender Cyberthreat Real-Time Map, Sonicwall Security Center (Worldwide Attacks), Akamai Real-Time Web Attack Monitor, LookingGlass Phishing/Malicious URL Map, Threat Butt Hacking Attack Map, Fortinet Threat Map, Talos Spam and Malware Map, Deteque Live Botnet Threats Worldwide and Sophos Threat Tracking Map. In the next section, the role of AI in real-time monitoring is explored.

5.2 Non-Threat Map-Based Monitoring Tools.

Monitoring is an activity that every organisation that is connected to the Internet should perform as part of normal business. In this section, we consider non- threat map-based monitoring tools as those that do not have or do not need to involve threat maps. Some of the well-known non-threat map-based monitoring tools are summarised in Table 1.

---

[18] https://www.digitalattackmap.com/#anim=1&color=0&country=ALL&list=0&time=18763& view=map.

[19] https://www.fireeye.com/cyber-map/threat-map.html.

Table 1 Non-threat map-based monitoring tools

Tool Name	Description
1. SolarWinds Server and Application Monitor[20]	This tool is designed to detect and troubleshoot performance issues. In addition, it can monitor applications and their supporting infrastructure, whether running on premises, in the cloud or a hybrid environment
2. Nagios Core[21]	Nagios Core, previously known as Nagios[22] is a tool designed to monitor systems, networks, and critical infrastructure. However, it also does monitoring and alerting services for servers, switches, applications, and services
3. VMware vRealize Operations[23]	VMware vRealize Operations is a unified operations platform that delivers continuous performance, capacity, and cost optimization, intelligent remediation and integrated compliance through AI/ML and predictive analytics
4. PRTG Network Monitor[24]	Developed by Paessler AG, PRTG Network Monitor is anagentless network monitoring tool used to monitor and classify system conditions like bandwidth usage or uptime as well as collect statistics and traffic from devices such as switches, routers, servers, and other applications in an organisation
5. InfluxDB[25]	Developed by InfluxData, InfluxDB is used to store and retrieve time series data in operations monitoring, application metrics, Internet of Things sensor data, real-time analytics among others. InfluxDB can handle massive volumes and countless sources of time-stamped data produced by sensors, applications, and infrastructure
6. Dynatrace[26]	Dynatrace provides a software intelligence platform based on artificial intelligence and automation to monitor and optimize applications. It can also support monitoring of infrastructure such as hosts, processes, and network including log monitoring, view total network traffic, hosts CPU usage, process response time and more

(continued)

---

[20] https://www.solarwinds.com/server-application-monitor.

[21] https://www.nagios.org/projects/nagios-core/.

[22] https://www.nagios.org/.

[23] https://www.vmware.com/au/products/vrealize-operations.html.

[24] https://www.paessler.com/prtg.

[25] https://www.influxdata.com/.

[26] https://www.dynatrace.com/.

Table 1 (continued)

Tool Name	Description
7. WhatsUp Gold[27]	WhatsUp Gold is a tool designed to provide easy visibility and management of device log data. WhatsUp Gold can monitor, filter, analyse, search, and provide alerts on syslogs or Windows logs for every device in a network while also watching for meta trends like log volume changes
8. Datadog[28]	Datadog is a monitoring and analytics tool (offering cloud Monitoring as a Service) that can be used to determine performance metrics as well as event monitoring for infrastructure and cloud-scale applications and services. It can also monitor services such as servers, databases, tools, and services, through a SaaS-based data analytics platform
9. Zabbix[29]	Zabbix is a monitoring tool designed for diverse IT environments, including networks, servers, cloud services, virtual machines, applications, and other services. Zabbix is flexible and scalable and supports distributed monitoring as well as other metrics, such as network utilisation, CPU load and disk space consumption
10. OPManager[30]	Developed by ManageEngine, OpManager is a real-time network monitoring tool designed to monitor availability, traffic, packet loss, performance metrics and response time of various devices such as servers, routers, switches and virtual machines, firewalls, load balancers, wireless LAN controllers, printers, storage devices, and everything that has an IP and is connected to the network. OpManager also does physical and virtual server monitoring and sends automatic notifications to relevant users via text message or email if any equipment goes down
11. SCOM[31]	Developed by Microsoft, System Center Operations Manager (SCOM) is a data centre monitoring system for operating systems and hypervisors. SCOM shows state, health, and performance information of computer systems as well as help individuals create, manage, monitor, and automate their infrastructure and workflows end-to-end

(continued)

---

[27] https://www.whatsupgold.com/.

[28] https://www.datadoghq.com/.

[29] https://www.zabbix.com/.

[30] https://www.manageengine.com/network-monitoring/.

[31] https://docs.microsoft.com/en-us/system-center/scom/welcome?view=sc-om-2019.

Table 1 (continued)

Tool Name	Description
12. VirtualWisdom[32]	VirtualWisdom® is Al-powered monitoring, analytics, and automation solution designed for IT Infrastructure Performance Monitoring (IPM) in production environments. VirtualWisdom® also allows enterprises to effectively monitor app-centric infrastructure performances. It also provides real-time as well as historical insights into availability, performance, utilisation, and health of data centres, including physical, cloud-based and virtual environments
13. New Relic[33]	New Relic is a tool designed to provide flexible, dynamic real-time server monitoring analytics from metrics, events, logs, and traces collected from a live production software, infrastructure, and services, transforming them into actionable insights about the entire stack, business, and customers' experiences
14. eG Enterprise[34]	Developed by eG Innovations, eG Enterprise is application performance and IT infrastructure monitoring platform that supports end-to-end IT performance monitoring and analytics. It provides automated IT monitoring, diagnosis, and reporting to troubleshoot application slowdowns, user experience issues, network outages, virtualisation bottlenecks, server failures, storage hotspots, and more
15. SiteScope[35]	Developed by Micro Focus, SiteScope is an agentless tool designed to support automated application monitoring as well as intuitive monitoring management. SiteScope can monitor the availability and performance of distributed IT infrastructure and applications including servers, operating systems, network services, virtualization software applications, and application components
16. LogicMonitor[36]	LogicMonitor is a fully automated, cloud-based infrastructure monitoring platform for enterprise IT and managed service providers. It can support full-stack visibility for networks, cloud, servers, and more within one unified view

(continued)

---

[32] https://www.virtana.com/products/virtualwisdom/.

[33] https://newrelic.com/platform.

[34] https://www.eginnovations.com/.

[35] https://www.microfocus.com/en-us/products/sitescope-application-monitoring/overview.

[36] https://www.logicmonitor.com/.

Table 1 (continued)

Tool Name	Description
17. Checkmk[37]	Checkmk is designed to monitor an entire hybrid IT Infrastructure and Applications which includes networks, servers, clouds, containers, the internet of things, databases, storage and applications
18. NetApp® OnCommand Insight[38]	NetApp® OnCommand Insight is designed for monitoring and troubleshooting an entire infrastructure in real-time from a single dashboard. The key components of OnCommand Insight are designed to help individuals better monitor, troubleshoot, optimize, and justify spending across an entire infrastructure. It allows you to know exactly how your systems are performing and being used
19. Applications Manager[39]	Developed by ManageEngine, the Applications Manager is an application monitoring system that provides deep visibility into the performance and user experience of business-critical applications and infrastructure components. It can also help proactively identify potential application problems and notify administrators when unexpected problems happen
20. Site24 × 7[40]	Site24 × 7 is a cloud-based platform that provides monitoring solutions to small and large business organisations. It brings together the monitoring of metrics, traces, and logs under one console for different layers of cloud architecture. Site24 × 7 also monitors websites, servers, clouds, networks, applications, and real-time users. This helps users to derive insights on uptime, web applications' performance, internet services and private networks
21. OpsRamp[41]	OpsRamp is a cloud-based IT operations management platform designed to monitor cloud and on-premises infrastructure as well as resolve incidents with machine learning. OpsRamp brings together on-premises and cloud infrastructure through discovery, monitoring, escalation management and automation
22. NinjaRMM[42]	NinjaRMM is a tool designed for remote monitoring and IT management solutions. NinjaRMM can offer, monitoring and alerting, endpoint, network, virtual machine, mobile app and other management services

(continued)

---

[37] https://checkmk.com/.

[38] https://www.netapp.com/data-management/oncommand-insight/.

[39] https://www.manageengine.com/products/applications_manager/.

[40] https://www.site24x7.com/.

[41] https://www.opsramp.com/the-opsramp-platform/.

[42] https://www.ninjarmm.com/.

Table 1 (continued)

Tool Name	Description
23. Sematext Cloud[43]	Sematext Cloud is a unified platform with all-in-one solutions for infrastructure, application performance, server, container, process, database, network, uptime, website, API, JVM, metrics and logs monitoring
24. Spiceworks Network Monitor[44]	Spiceworks Network Monitor is designed for monitoring servers and switches. It can also be used to create and view alerts, track disk space, and monitor the health of the Microsoft exchange server. It can also get detailed stats including hardware configuration, installed OS, software, patches, IP address and more for each device
25. Atera[45]	Atera is an IT management solution that enables remote monitoring, management and automation of IT networks which includes remote control, patch management, discovery, inventory of IT assets, monitoring, security, backup, and more
26. Cacti[46]	Cacti is a web-based network monitoring and graphing tool as well as a robust and extensible operational monitoring and fault management framework
27. Icinga[47]	Icinga is a computer system and network monitoring application with support for infrastructure, automation, and cloud monitoring
28. OpenNMS[48]	OpenNMS is an enterprise-grade network monitoring and network management platform
29. AppDynamics[49]	AppDynamics is an application performance management and IT operations analytics platform designed to help build dashboards that automatically correlate application performance to business, user, applications, infrastructure, network, and security outcomes
30. Pingdom[50]	Pingdom is a website performance and availability monitoring platform that includes end-user experience monitoring. It is also used to report whether a website is down due to network splits or failure in DNS servers
31. mPulse[51]	Designed by Akamai, mPulse is a real user monitoring (RUM) solution that enables companies to monitor, find, and fix website and application performance issues

(continued)

---

[43] https://sematext.com/cloud/.

[44] https://www.spiceworks.com/free-network-monitoring-management-software/.

[45] https://www.atera.com/.

[46] https://www.cacti.net/.

[47] https://icinga.com/.

[48] https://www.opennms.com/.

[49] https://www.appdynamics.com/company/why-appdynamics.

[50] https://www.pingdom.com/.

[51] https://www.akamai.com/products/mpulse-real-user-monitoring.

Table 1 (continued)

Tool Name	Description
32. Raygun[52]	Raygun is a Real User Monitoring tool for web and mobile applications. It is used to track website and application performance, quickly identify, and resolve issues, and improve customers' experience
33. Atatus[53]	Atatus is a SaaS-delivered application performance and error tracking solution which includes application performance Monitoring, Real User Monitoring, Infrastructure Monitoring, Database Monitoring and Log Management & Analytics

6   Role of AI in Real-Time Monitoring of Cyber Security Threats

Having explored the different cybersecurity threats as well as common cybersecurity vulnerabilities, this section is dedicated to explaining how AI can be used to enhance the real-time monitoring of cybersecurity threats. This is important because as the demand for more explainable AI models with interpretable predictions rises, so does the need for AI models that can aid in achieving real-time monitoring of cybersecurity threats.

Real-time monitoring combined with AI, specifically unsupervised machine learning algorithms, can "learn" the access behaviour of remote machinery and equipment, and determine the probability of theft or breach.

6.1   Event or Incident Prediction

Artificial Intelligence, and specifically unsupervised machine learning algorithms, can be used together with real-time monitoring tools to help "learn" for example, access behaviours of users. This way organisations can use real-time data and analytics as well as machine learning models to predict the probability of an unauthorised intrusion to a remote machine or equipment. Having these predictive insights powered by AI, effective deterrence strategies can be launched to protect devices and properties. A mixture of AI, real-time monitoring, and continuous feeds of data from monitoring can reduce false alarms and help monitoring teams to be more effective [50].

---

[52] https://raygun.com/platform/real-user-monitoring.

[53] https://www.atatus.com/

6.2 Location Intelligence and Situational Awareness

Artificial Intelligence can be used to enable location intelligence and situational awareness. Moreover, with machine learning, risk scores can be updated in real-time to help to prevent unauthorised access to different systems or physical locations. AI can also help eliminate compromised credential attacks by learning in real-time users' behaviours, patterns, context, and device usage and using the data to verify trusted user access [50]. This use case is important in the medical IoT domain, with the growing number of implantable medical devices that may also require authentication [51].

6.3 Effective Zero-Day Threat Detection

Artificial Intelligence can help detect zero-day malware. By using machine learning algorithms, for example, AI systems can be trained to detect malware, run pattern recognition, and detect even the minutest behaviours of malware or ransomware attacks in real-time before they enter an organisation network or system [52]. AI can also give organisations intelligence of new anomalies, cyberattacks as well as prevention strategies.

6.4 Advanced Endpoint Protection

Endpoint protection aims to defend critical data found on endpoint devices, such as laptops, smartphones, tablets among others. With the emergence of the Internet of Things (IoT) and the recent COVID-19 pandemic, many endpoint devices are used for working from home. AI can be used to detect threats in real-time by applying pattern recognition and machine learning to inspect endpoint devices. AI can also learn the features of a compromised file. AI-driven endpoint protection can also be used to establish a baseline of behaviour for different endpoints through a repeated training process and if outliers are detected, AI can flag them and act. This further, provides proactive protection against threats as well as monitors and secure endpoints in real-time [52]. Machine learning can also be used to derive risk scores based on previous behavioural patterns, geolocation, time of login, and other variables thus helping in securing and controlling access to endpoints.

6.5   Enhanced Vulnerability Management

Vulnerability management is a process of identifying, evaluating, treating, and reporting on security vulnerabilities in systems and the software that runs on them. Modern vulnerability management systems can embrace AI to enhance vulnerability scans in an organisation's IT infrastructure as well as help in faster incidence responses to decrease risks. With the growing number of software applications being used in different organisation departments, AI can further be used to flag in real-time vulnerabilities that criminals would typically exploit. However, AI can also be used to enhance existing detection tools to better pinpoint weaknesses in user devices before a threat occurs [53].

6.6   Effective Security Recommendations

AI can simplify data analytics to users who do not have a technical background thus making it easy for them to understand cyber risks within an organisation. Moreover, with advanced AI technologies, it is also possible to get more specialised security recommendations that can help in the mitigation process. Detailed and effective explanation further makes it easy for users to understand existing issues which in return helps them to be more aware of how to mitigate future security incidents or vulnerabilities. [53].

7   Conclusions and Future Work

With the speed of technological change, we believe that there is a clear need for advanced automated real-time monitoring and detection systems, informed by sophisticated Artificial Intelligence algorithms: not only to anticipate, recognise, detect, and mitigate novel attacks but also to avoid wasting time and resources with false alarms. Many areas of cybersecurity will require more research to improve the current monitoring and detection systems. These include developing monitoring and detection systems with the ability to handle challenges in real-time and more especially with a high detection ratio and performance. Further, exploring new techniques of fast incremental learning could be useful for the real-time update of models for anomaly detection. Also, developing new techniques or algorithms that can detect new incoming threats and respond automatically without intervention from a human administrator and developing flexible systems that can self-orchestrate a security policy is a rich area for research. Finally, developing intelligent systems that have a centralised view of the network security of an organisation is paramount as complex systems do interact and treating systems as disconnected entities and not part of a whole cyber ecosystem is no longer an option.

References

1. ISF, "Disruption. Distortion. Deterioration," securityforum, 2019. https://www.securityforum. org/research/threat-horizon-2019/. [Accessed 2 Sept. 2019]
2. V.R. Kebande, N.M. Karie, R.A. Ikuesan, Real-time monitoring as a supplementary security component of vigilantism in modern network environments. Int. J. Inf. Technol. **13**(1), 5–17 (2021)
3. P. Linardatos, V. Papastefanopoulos, S. Kotsiantis, Explainable ai: A review of machine learning interpretability methods. Entropy **23**(1), 1–45 (2021)
4. D. Gunning, D. Aha, DARPA's explainable artificial intelligence (XAI) program. AI Mag. **40**(2), 44–58 (2019)
5. RoyalSociety, *Explainable AI,* "*The Royal Society*, 28 November 2019. https://royalsociety. org/topics-policy/projects/explainable-ai/. [Accessed 04 Nov. 2021]
6. N.J. Nilsson, *The Quest for Artificial Intelligence* (Cambridge University Press, Cambridge, 2009)
7. H. Taylor, *What Are Cyber Threats and What to Do About Them*, preyproject, 16 June 2021. https://preyproject.com/blog/en/what-are-cyber-threats-how-they-affect-you-what-to-do-about-them/. [Accessed 05 Aug. 2021]
8. N. M. Karie, *Resolving Semantic Disparities in Digital Forensics* (2016). https://repository.up. ac.za/handle/2263/57356. [Accessed 05 Aug. 2021]
9. J. Vukalović, D. Delija, Advanced persistent threats-detection and defense, in *38th International Convention on Information and Communication Technology, Electronics and Microelectronics (MIPRO)* (Opatija, Croatia, 2015)
10. N.M. Karie, V.R. Kebande, H.S. Venter, Diverging deep learning cognitive computing techniques into cyber forensics. Forens. Sci. Int.: Synergy **1**, 61–67 (2019)
11. A. DeNisco-Rayome, *Five Emerging Cybersecurity Threats You Should Take Very Seriously in 2019*, ZDNET, 15 February 2019. https://www.zdnet.com/article/five-emerging-cybersecu rity-threats-you-should-take-very-seriously-in-2019/. [Accessed 05 Aug. 2021]
12. IC3, *What is Ransomware?* FBI, Internet Crime Compliant Center IC3, 2021. https://www.ic3. gov/Home/Ransomware. [Accessed 05 Aug. 2021]
13. Crowdstrike, *Types of Ransomware*, Crowdstrike, 27 May 2021. https://www.crowdstrike.com/ cybersecurity-101/ransomware/types-of-ransomware/. [Accessed 25 Aug. 2021]
14. A. Skrop, DATALEAK: data leakage detection system. Sciendo **1**(1), 113–124 (2015)
15. Forcepoint, *What is Data Leakage? Data Leakage Defined, Explained, and Explored*, Forcepoint, 2021. https://www.forcepoint.com/cyber-edu/data-leakage. [Accessed 05 Aug. 2021]
16. Y. Lu, L.D. Xu, Internet of things (IoT) cybersecurity research: a review of current research topics. IEEE Internet Things J. **6**(2), 2103–2115 (2018)
17. N.A.G. Arachchilage, S. Love, K. Beznosov, Phishing threat avoidance behaviour: an empirical investigation. Comput. Hum. Behav. **60**, 185–197 (2016)
18. I. Vayansky, S. Kumar, Phishing – challenges and solutions. Comput. Fraud Secur. **2018**(1), 15–20 (2018)
19. S. McCombie, J. Pieprzyk, Winning the phishing war: a strategy for Australia, in *2010 Second Cybercrime and Trustworthy Computing Workshop,* Ballarat (2010)
20. APWG, *Phishing Activity Trends Report*, APWG, 2021. https://apwg.org/trendsreports/. [Accessed 05 Aug. 2021]
21. F. Mouton, L. Leenen, H.S. Venter, Social engineering attack examples, templates and scenarios. Comput. Secur. **59**, 186–209 (2016)
22. S. Kumar, Hacking attacks, methods, techniques and their protection measures. Int. J. Adv. Res. Comput. Sci. Manag. **4**, 2353–2358 (2018)
23. A. Ushmani, Ethical hacking. Int. J. Inform. Technol. (IJIT) **4**(6) (2019)
24. T. Noonan, E. Archuleta, National Infrastructure Advisory Council's Final Report and Recommendations on the Insider Threat to Critical Infrastructures, 2008. [Washington, D.C.] : DHS/NIAC (2008) . [Accessed 05 Aug. 2021].

25. Ekran, 5 real-life examples of breaches caused by insider threats. Ekransystem, 18 November 2020. https://www.ekransystem.com/en/blog/real-life-examples-insider-threat-cau sed-breaches. [Accessed 25 Aug. 2021]

26. G. Carl, G. Kesidis, R.R. Brooks, S. Rai, Denial-of-service attack-detection techniques. IEEE Internet Comput. **10**(1), 82–89 (2006)

27. F. Lau, S.H. Rubin, M.H. Smith, L. Trajkovic, Distributed denial of service attacks, in *2000 IEEE International Conference on Systems, Man and Cybernetics. 'Cybernetics Evolving to Systems, Humans, Organisations, and Their Complex Interactions* (Nashville, 2000)

28. Malwarebytes, *Cryptojacking – What is it?*, Malwarebytes, 2020. https://www.malwarebytes. com/cryptojacking. [Accessed 06 Aug. 2021]

29. J.E. Scheuermann, *Cyber-Physical Attacks On Critical Infrastructure: What's Keeping Your Insurer Awake At Night?*, K&L Gates LLP, 24 January 2017. https://www.klgates.com/cyber-physical-attacks-on-critical-infrastructure--whats-keeping-your-insurer-awake-at-night-01-24-2017. [Accessed 06 Aug. 2021]

30. G. Murray, M.N. Johnstone, C. Valli, The convergence of IT and OT in critical infrastructure, in *15th Australian Information Security Management Conference* (Perth, 2017)

31. Paloalto, What is an endpoint?. Paloalto Netw. (2020). https://www.paloaltonetworks.com/cyb erpedia/what-is-an-endpoint. [Accessed 06 Aug. 2021]

32. S. Droski, *Endpoint Guide: Threats, Security, Visibility, and Protection*, Secureworks, 29 June 2021. https://www.secureworks.com/blog/endpoint-security-guide-management-protec tion-detection. [Accessed 06 Aug. 2021].

33. Z.A. Baig, P. Szewczyk, C. Valli, P. Rabadia, P. Hannay, M. Chernyshev, M. Johnstone, P. Kerai, A. Ibrahim, K. Sansurooah, N. Syed, M. Peacock, Future challenges for smart cities: Cyber-security and digital forensics. Digit. Investig. **22**, 3–13 (2017)

34. N. Karie, N. Sahri, P. Haskell-Dowland, IoT threat detection advances, challenges and future directions, in *Workshop on Emerging Technologies for Security in IoT (ETSecIoT 2020)* (Sydney, 2020)

35. A. Woodiss-Field, M.N. Johnstone, Assessing the suitability of traditional botnet detection against contemporary threats, in *Workshop on Emerging Technologies for Security in IoT (ETSecIoT 2020)* (Sydney, 2020)

36. S. Robinson, *Smart Home Attacks Are A Reality, Even As The Smart Home Market Soars* (Cisco, 2019). https://www.cisco.com/c/en/us/solutions/internet-of-things/smart-home-attacks.html. [Accessed 10 Sept. 2019]

37. A. Atkinson, *Cybersecurity Tech Basics: Vulnerability Management: Overview*, July 2018. [Online]. Available: https://www.cisecurity.org/wp-content/uploads/2018/07/Cybersecurity-Tech-Basics-Vulnerability-Management-Overview.pdf. [Accessed 06 Aug. 2021]

38. J.M. Kizza, *Guide to Computer Network Security, Tennessee, Chattanooga* (Springer International Publishing AG, USA, 2017)

39. Owasp, *Using a Broken or Risky Cryptographic Algorithm*, Open Web Application Security Project, 2020. https://owasp.org/www-community/vulnerabilities/Using_a_broken_or_risky_cryptographic_algorithm. [Accessed 06 Aug. 2021]

40. Martello, *Use of a Broken or Risky Cryptographic Algorithm*, Martello Security, 2021. https://www.martellosecurity.com/kb/mitre/cwe/327/. [Accessed 06 Aug. 2021]

41. B. Eshete, A. Villafiorita, K. Weldemariam, Early detection of security misconfiguration vulnerabilities in web applications., in *Sixth International Conference on Availability, Reliability and Security* (Vienna, Austria, 2011)

42. B. Martin, M. Brown, A. Paller, D. Kirby, S. Christey, "CWE/SANS top 25 most dangerous software errors. http://cwe.mitre.org/top25/archive/2010/2010_cwe_sans_top25.pdf. [Accessed 05 Nov. 2021]

43. B. Grobauer, T. Walloschek, E. Stocker, Understanding cloud computing vulnerabilities. IEEE Secur. Priv. **9**(2), 50–57 (2010)

44. T. Nafees, N. Coull, I. Ferguson, A. Sampson, Vulnerability anti-patterns: a timeless way to capture poor software practices (vulnerabilities), in *24th Conference on Pattern Languages of Programs* (Vancouver, Canada, 2018)

45. M.D. Kaur, P. Kaur, Secure web development based on vulnerabilities. Int. J. Electron. Commun. Comput. Eng. **3**(1), 120–125 (2012)
46. I. Broderick, *Insecure Cryptographic Storage Explained*, Veracode, https://www.veracode.com/blog/2012/06/insecure-cryptographic-storage-explained. [Accessed 06 August 2021].
47. T.A. Nidecki, *What Are Insecure Direct Object References*, Acunetix, https://www.acunetix.com/blog/web-security-zone/what-are-insecure-direct-object-references/. [Accessed 06 Aug 2021]
48. Owasp, *Unrestricted File Upload*, OWASP, https://owasp.org/www-community/vulnerabilities/Unrestricted_File_Upload. [Accessed 06 Aug 2021]
49. Secureflag, *Unrestricted File Upload*, Secureflag, https://knowledge-base.secureflag.com/vulnerabilities/unrestricted_file_upload/unrestricted_file_upload_vulnerability.html. [Accessed 06 August 2021].
50. L. Columbus, *10 Ways AI & Real-Time Monitoring Can Protect Utilities Today*, Forbes, 11 July 2020. https://www.forbes.com/sites/louiscolumbus/2020/06/11/10-ways-ai--real-time-monitoring-can-protect-utilities-today/?sh=73af8de34b91. [Accessed 06 Aug 2021]
51. G. Zheng, W. Yang, C. Valli, L. Qiao, R. Shankaran, M.A. Orgun, S.C. Mukhopadhyay, Finger-to-heart (F2H): authentication for wireless implantable medical devices. IEEE J. Biomed. Health Inform. **23**(4), 1546–1557 (2018)
52. G. Belani, *The Use of Artificial Intelligence in Cybersecurity: A Review*, IEEE. https://www.computer.org/publications/tech-news/trends/the-use-of-artificial-intelligence-in-cybersecurity. [Accessed 06 Aug 2021]
53. V. Varadaraj, *The What, Why, and How of AI and Threat Detection*, McAfee, 02 June 2021. https://www.mcafee.com/blogs/consumer/consumer-cyber-awareness/the-what-why-and-how-of-ai-and-threat-detection/. [Accessed 06 August 2021].

Chapter 8
Network Forensics in the Era of Artificial Intelligence

Wencheng Yang, Michael N. Johnstone, Song Wang, Nickson M. Karie,
Nor Masri bin Sahri, and James Jin Kang

Abstract Network forensics investigates a network attack by tracing the source of the attack and attributing the crime to a person, host or network. It can antici-pate prospective attacks by establishing attack patterns based on available evidence and intrusion data traces. This chapter introduces network forensics, describes some common attacks targeting networks and existing network forensic tools. Moreover, this chapter describes the current development of network forensics techniques, such as IP Traceback Techniques, Intrusion Detection Systems, Attack Graph-based Tech-niques, Honeypots and Privacy-preserving Data Analytics. Based on the above, some specific research gaps in current network forensics research in the era of artificial intelligence are identified.

Keywords Network frensics · Cyberattacks · Intrusion detection systems ·
Artificial intelligence

W. Yang (✉) · M. N. Johnstone · N. M. Karie · N. M. Sahri · J. J. Kang
School of Science, Cyber Security Cooperative Research Centre, Edith Cowan University,
Joondalup, Australia
e-mail: w.yang@ecu.edu.au

M. N. Johnstone
e-mail: m.johnstone@ecu.edu.au

N. M. Karie
e-mail: n.karie@ecu.edu.au

N. M. Sahri
e-mail: n.sahri@ecu.edu.au

J. J. Kang
e-mail: james.kang@ecu.edu.au

S. Wang
School of Engineering and Mathematical Sciences, La Trobe University, Melbourne, Victoria,
Australia
e-mail: song.wang@latrobe.edu.au

© The Author(s), under exclusive license to Springer Nature Switzerland AG 2022
M. Ahmed et al. (eds.), *Explainable Artificial Intelligence for Cyber Security*,
Studies in Computational Intelligence 1025,
https://doi.org/10.1007/978-3-030-96630-0_8

1   Introduction

Network forensics, also known as digital forensic science in networked environments, is a subfield of digital forensics that discovers and retrieves evidence-related information about crimes in a network environment to make it acceptable in court. [1]. Various researchers have more comprehensive definitions of network forensics, for example, Palmer in [2] states: *"The use of scientifically proven techniques to collect, fuse, identify, examine, correlate, analyse, and document digital evidence from multiple, actively processing and transmitting digital sources for the purpose of uncovering facts related to the planned intent, or measured success of unauthorized activities meant to disrupt, corrupt, and or compromise system components as well as providing information to assist in response to or recovery from these activities"*.

Digital forensics focuses on recovering and investigating digital materials or evidences, such as document and image files, often in relation to computer crimes [3]. There are several considerations in the courtroom when it comes to digital evidences [4]. The two main standards of digital evidence regarding the admissibility of expert witness testimony are the Frye standard and the Daubert standard. The former requires rigorous scientific grounding from testing, peer review, and error rates to be acceptable in the relevant scientific community. In contrast, the Daubert standard is relatively more relaxed yet still ensures that the processes and methods used during an investigation process are backed by science (and is the test used in courts in USA). However, even this can open challenges such as contesting the qualifications of associated knowledgeable experts, examining the scientific nature of the proof and the validity/dependability of the hardware and code tools (e.g., write blockers and code suites) utilised [5]. Clearly, recourse to legal action is not the only goal or motivation for a network forensics exercise – a firm may simply wish to understand the nature of an incursion, leading to an improvement in network security, thus any legal requirement might be secondary to the primary goal of regaining control of, and securing, an asset. Nonetheless, the techniques used in digital forensics ought to satisfy legal and technical requirements, viz.[1]:

1.   Rules of evidence [6]:

- Admissible. Can be used in court or elsewhere that the evidence is needed.
- Authentic. Evidence should be related to an incident in a relevant way.
- Complete. No tunnel vision and exculpatory evidence for alternate suspects.
- Reliable. The evidence should be authentic and integral.
- Believable. The evidence should be clear and easy to understand, and a jury should believe it.

---

[1] https://www.sciencedirect.com/topics/computer-science/network-forensics.

2.   Criteria for admissibility of novel scientific evidence/expert testimony (Daubert criteria; standardised in U.S. Federal law) [6]:

- The theory or technique should be tested reliably.
- The theory or technique should be peer-reviewed and published.
- The error rate of the method used should be clear.
- The theory or method should have been generally accepted by the scientific community.

In the early days (about 40 to 50 years ago), digital forensics techniques were primarily focused on data recovery. For example, in the vignette related by Wood et al., two data recovery consultants worked for several days to recover a replica of a terribly fragmented database file deleted by a careless researcher [7]. Later, from 1999 to 2007, digital forensics became a "magical window" that could see into a) the past, by restoring residual data that was deemed to have been deleted; and b) criminal minds, by restoring emails and instant messages. Moreover, network and memory forensics can effectively snapshot time and observe criminal behaviours even months after an incident [8]. Nowadays, the crisis facing digital forensics is attributed to the rapid rate of progress in the computer industry [8]; for example, the size of storage devices is dramatically increasing. Compounding the problem for network forensics is the advent of smart cities [9], where there exist a plethora of heterogeneous networks, protocols and devices, coupled with non-standard connections and usage patterns brought on by the COVID-19 pandemic. Forensics cases face challenges such as the need to analyse multiple devices and the use of the cloud for remote sensing, processing and storage.

Digital forensics investigations are not limited to retrieving data from computers, as end-users now have small digital devices such as smartphones, tablets and flash drives, some of which have volatile memory whilst others may have non-volatile memory. Digital forensics investigation can be split into the following types, depending on the type of device, media or artifact [10, 11]:

- Computer forensics [12]
- Network forensics [13]
- Mobile/smartphone forensics [14]
- Database forensics [15]
- Others (e.g., memory forensics [16], email forensics [17], social media forensics [18], cloud forensics [19], Internet of Things (IoT) forensics [20])

Unlike other areas of digital forensics that focus on storage of static data, network forensics handles volatile and dynamic data, which opens a unique set of challenges in terms of the chain of custody. Network forensics often has two focal points:

- First, it is related to security and involves detecting abnormal traffic and identifying intrusions.
- Second, it is related to law enforcement involving capturing and analysing network traffic.

The evolution of cybersecurity and its associated forensic processes and related toolsets are largely driven by the latest advances in Internet technology. As more aspects of our daily lives are migrating to online systems and databases, to which threats of criminal activity are increasing, so the demand for sophisticated analytical tools is also increasing [13]. There are some common reasons for using network forensics, such as analysing computer systems; gathering evidence for use in a court; recovering data in the event of a hardware or software failure; and collecting and analysing live data packets to detect malicious attacks [13].

The structure of this chapter is organized as follows. Section 2 introduces a typical process model of network forensics. Section 3 lists some common types of network attacks. Section 4 presents tools for network forensics. The research advances and research gaps in network forensics research are presented in Sects. 5 and 6, respectively. In Sect. 7, the explainable artificial intelligence for network forensics is discussed.

2   Process Model of Network Forensics

In general, there are three main steps in network forensics according to [21], viz., capture, identify and analyse.

- Capture: This step uses standard methods and procedures to record the physical scene and replicate digital evidence.
- Identify: This step is based on the network indicators to identify and determine the event.
- Analyse: This step is to determine the importance and rebuild the network traffic packet and draw conclusions based on the evidence found.

A detailed process model of network forensics is presented in [22]. This process model is summarised in Fig. 1. A brief overview of each step is provided below.

- Preparation: deployment of network security tools, e.g., a packet analyser and intrusion detection systems (IDSs), at different points of the network.
- Detection: Observation of alerts created by those security tools deployed at the network and a quick assessment and confirmation of the suspected attack.
- Incident response: A response is launched based on the information gathered on an alleged crime or detected intrusion.
- Collection: Traffic is collected from the network. This phase is very important because of the change of data, and it is impossible to generate precisely the same trace at a later time.
- Preservation: The obtained network traffic data are stored as backups in the form of traces, logs, etc. A copy of the data will be analyzed while the original data are left untouched.
- Examination: The traces obtained are integrated and fused into a large dataset. Then analysis can be performed on this dataset.

Fig. 1 A generic process
model for network forensics
[22]

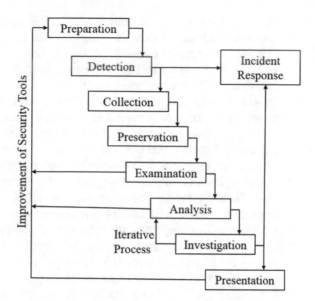

- Analysis: Use statistical, data mining and soft computing methods to deduce important observations based on existing attack patterns.
- Investigation: The aim of the investigation step is to determine the point of attack origin.
- Presentation: Observations are presented in a language understandable by the desired audience. An explanation is provided with various procedures used to draw the conclusions.

3   Types of Network Attacks

An attack can be launched from various machines connected to the network either via wired or wireless media (given the interconnectivity of the Internet, here, the "network" can mean any local or remote connection using a range of protocols). Network attacks are launched every hour and are growing at an alarming rate.[2] Some attacks are briefly summarised below.

- Browser Attacks [23]: A browser attack attempts to infiltrate a computer via a web browser, which is one of the most common ways that people access the Internet. A browser attack usually starts on legitimate, but vulnerable, websites. An attacker can destroy a site or infect it with malware. When a new visitor arrives through a web browser, the infected site attempts to exploit the vulnerability in the guest browser to force malware into the visitor's system.

---

[2] https://aag-it.com/how-often-do-cyber-attacks-occur/.

- Brute Force Attacks [23]: A brute force attack is analogous to breaking down the front door of a house. Instead of trying to trick a user into downloading malware, the attacker attempts to determine the authorisation credentials of the service or system by trial and error. This could be achieved manually or with the aid of open-source software such as Hashcat.
- Malware Attacks [23]: The uses of malware have been extensively disseminated and are widely known. Three common ways of spreading malware include: (1) Phishing emails—malicious attackers generate messages to lull victims into a false sense of security and entice them to download attachments that are actually malware. (2) Malicious websites—malicious attackers can create websites that contain exploit kits which are designed to discover vulnerabilities in the systems of site visitors (victims) and force them to embed malware into their systems. These sites could also be utilized to disguise malware as legitimate downloads. (3) Malvertising—malicious attackers have discovered ways to distribute their products using advertising networks. When clicked, malicious advertisements can redirect a user to a website hosting malware. Some malicious attacks can infect a system without requiring user interaction.
- Worm Attacks [23]: Malware usually needs user interaction to initiate an infection. For instance, a user may download an email attachment that is malicious, access a website that is infected, or insert an infected flash drive into a computer. In contrast, a worm attack spreads on its own. Malware that does not require user interaction is known as self-propagating malware. Often, by exploiting system vulnerabilities, such malware can spread to local networks and elsewhere.
- Denial of Service (DoS) attacks [23]: A DoS attack is a commonly found, yet serious, class of attack. It attempts to temporarily or indefinitely interrupt the services of a host connected to the Internet, rendering the target user unable to use machine or network resources. Denial of Service is usually accomplished by injecting redundant requests into the target machine or resource to overload the system and prevent some or all legitimate requests from being fulfilled.
- Botnet [24]: This type of attack refers to a private computer network that is infected by malware and controlled as a group without the owner's knowledge. The network engages in certain activities, such as sending spam, malware or being a slave (zombie) in a DoS attack.
- IP Spoofing [25]: This attack is designed to inject packets with the wrong source address into the Internet. This is just one of many ways in which a user can pretend to be another user.

New attacks are detected every year. Figure 2 indicates the top attack types in 2019 versus 2020 demonstrated in percentage of total attacks recorded by IBM.

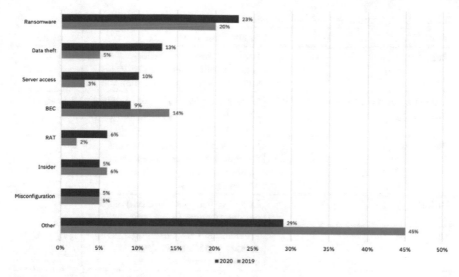

Fig. 2 Top attack types in 2019 versus 2020 (*Source* IBM Security X-Force)[3]

4   Tools for Network Forensics

Table 1 lists some common tools for conducting network forensics investigations. In practice, it is rarely possible to conduct any significant investigation with a single tool, due to the aforementioned heterogeneity of connected devices, protocols and storage options. Typically, investigators need to combine multiple tools [13].

The following is a summary of the functionality of some of the tools in Table 1.

Xplico[4]: As an open-source network forensic analysis tool, Xplico can capture the application data contained in Internet traffic. For instance, Xplico can extract email, HTTP content and FTP content from a pcap file.

Wireshark[5]: Wireshark is a free, open-source packet analyser. It can be used for a variety of purposes such as network troubleshooting, software and communications protocol development, analysis and education.

NetworkMiner[6]: NetworkMiner as an open-source network forensic analysis tool can be used for Windows, Linux, Mac OS X or FreeBSD. It can be utilized as a passive network sniffer or packet capture tool to detect operating systems, open ports and so on without having to place any traffic on the network.

---

[3] https://www.ibm.com/downloads/cas/M1X3B7QG.

[4] https://www.xplico.org/.

[5] https://en.wikipedia.org/wiki/Wireshark.

[6] https://www.netresec.com/?page=NetworkMiner.

Table 1 Some commonly used tools for network forensics investigations (*Source* [13])

Tool	Features and Advantages	Attributes*
Argus	Used for network forensics, nonrepudiation, detecting very slow scans, and supporting zero-day attacks	F, L
Airmon-ng, Airodumpng, Aireplay-ng, Aircrack-ng	Widely used suite of low-level traffic analysis tools for wireless LANs; used in Backtrack v5	F, L, R, C
DeepNines	Provides real-time identity-based network defense for content and applications, along with basic network forensics	F
Driftnet	Listens to network traffic and picks out images; used in Backtrack v5	F
Dragon IDS	Provides network, host intrusion detection and network forensic capture analysis	F, R, L, C
EtherApe	Graphical network monitor for capturing network traffic	F
Fenris	Suite of tools for code analysis, debugging, protocol analysis, reverse engineering, network forensics, diagnostics, security audits, vulnerability research	F
Flow-Tools	Software package for collecting and processing NetFlow data from Cisco and Juniper routers	F, L
Forensic and Log Analysis GUI	Log file analysis combined with network forensics; Python implementation	L
Honeyd	Improves cybersecurity by providing mechanisms for traffic monitoring, threat detection, and assessment	F
Infinistream, nGenius	Appliance for network forensics, incident analysis combined with session reconstruction and playback	F, R, C
Kismet	Network detector, network packet sniffer, and intrusion-detection system for wireless LANs	F
Ngrep	Simple, low-level network traffic debugging tool	F
NetworkMiner	Network forensic analysis tool that can be used as a passive network sniffer/packet-capturing tool	F

(continued)

Table 1 (continued)

Tool	Features and Advantages	Attributes*
NetStumbler	Widely used wireless LAN analysis tool for devices and network traffic analysis	F
NetDetector	Appliance for network forensic analysis, network security surveillance, signature-based anomaly detection	F, R, C, A
NetIntercept	Appliance for network forensics, monitoring, and analysis	F, R, C, A
NetWitness	Addresses network forensic analysis, insider threat, data leakage protection, compliance verification, designer malware, and 0-day detection	F, L, R, C, A
Omnipeek, Etherpeek	Low-level traffic analyzer for network forensics	F, L, R
RSA EnVision	Provides live network forensics analysis, log management, network security surveillance, data leakage protection	F, L, R, C, A
Snort	Widely used, popular tool for network intrusion detection and prevention, as well as for network forensic analysis	F
Savant	Appliance for live forensic analysis, surveillance, network analysis, and critical infrastructure reporting	F, R
Solera DS	Appliance for live network forensics, application classification, metadata extraction, and analysis tools	F, R, C, A
TCPDump, Windump	Command-line network packet analyzer that supports network forensic analysis	F
Wireshark	Widely used network traffic analysis tool; forms basis of network forensics studies	F
Xplico	Network forensic analysis tool that allows for data extraction from traffic captures; used in Backtrack v5	F

*F: filter and collect; L: log analysis; R: reassembly of data stream; C: correlation of data; A: application-layer view

Table 2 Some popular AI-driven tools/products

Tools/Products	Features	Source
Darktrace Antigena Network	An autonomous response solution for the enterprise. Powered by self-learning AI; can interrupt attacks at machine speed	https://www.darktrace.com/en/products/antigena-network/
SolarWinds Security Event Manager	Lightweight, ready-to-use; security information and event management product	https://www.solarwinds.com/security-event-manager
IBM Qradar	Accurately detect and prioritize threats; available on premises and in a cloud environment	https://www.ibm.com/products/qradar-siem
Suricata	Fast and robust network threat detection engine; real time intrusion detection (IDS), inline intrusion prevention (IPS), network security monitoring (NSM)	https://suricata-ids.org/
AlienVault (AT&T Cybersecurity)	Asset discovery, Vulnerability assessment, Intrusion detection, Behavioral monitoring	https://cybersecurity.att.com/products/ossim

TCPDump[7]: A free (BSD license) packet analyser, TCPDump runs on the command line. It displays TCP/IP frames being sent or received by the network to which the computer is connected.

Snort[8]: Snort is a network-based intrusion detection and prevention system. By performing protocol analysis, content search and matching, it enables real-time traffic analysis and packet logging on the network. It can also be utilized for detecting probes or attacks, such as semantic URL attacks and buffer overflows. Snort consists of three main modes, namely sniffer, packet logger and network intrusion detection.

The tools listed in Table 1 work with different network layers and operate at different levels to produce their output. Thus, Xplico operates quite differently to TCPDump. As mentioned previously, the problem space has increased because of the sheer volume of data being transmitted across networks, which leads to a "perfect storm" considering the number of heterogeneous networks, protocols and devices. This means that aggregators and dashboards driven by artificial intelligence (AI) have become popular, but that does not mean that a low level has fallen into disuse. Table 2 shows some popular AI-driven tools/products for network security.

---

[7] https://en.wikipedia.org/wiki/Tcpdump.

[8] https://en.wikipedia.org/wiki/Snort_(software).

5   Research Advances in Network Forensics

5.1   IP Traceback Techniques

In network forensics, the identification of the source of packets is called IP traceback. IP traceback techniques can be utilised to identify the source of the generated packet by identifying the source of the attack and by investigating attack paths, especially for distributed denial of service (DDoS) and IP spoofing attacks. To keep the network system secured, it is necessary to incorporate various traceback mechanisms/methods in an efficient manner. In [26], a deterministic packet marking (DPM) scheme for preventing DDoS or conventional DoS attacks using an IP traceback technique is proposed. In the scheme, a hash of the IP address of an ingress router is pre-calculated and split into several segments. When marking a packet, the router randomly selects a segment to mark into the packet. The victim uses the map of its upstream routers to identify the marked router in the traceback stage. According to the map, the victim can identify the candidate ingress router by receiving only a small number of marked packets. Fen et al. [27] proposed a lightweight, efficient but robust IP traceback scheme by recording TTL of the IP header of packets to trace the attack path. The proposed scheme can be used at any strength of DDoS attacks. In [28], Jeong and Lee designed an IP traceback protocol, which consists of a compressed hash table, a sinkhole router and data mining based network forensics, to defend network attacks. The proposed protocol can trace an attack in real time utilizing a hash table. Also, this protocol can detect a replay attack using timestamps to messages to verify integrity.

5.2   Intrusion Detection Systems

An intrusion detection system protects a network and monitors the malicious activities of an attacker on the network when the attacker tries to compromise the network. Packet inspection is a common and popular method to detect anomaly and attacks. Whilst traditional port matching has been widely used to identify targeted packets, perform deep packet inspection (DPI), inspects headers and payload unencrypted to identify the originating network applications. It ensures data to be in a correct format, and in the case of malicious codes, actions to block, re-route and log events. Port mirroring, also called Span Port, is a common way to acquire packets [29]. Based on the detection of an attack, the intrusion detection system can trigger an alert message and inform another system, such as a security information and event management (SIEM) system. Generally, logging approaches are used in intrusion detection systems to describe forensic explanations and reliability of evidence and analyse network intrusion data and so on [30]. Some of the methods to identify breaches found in networks are described in the following. In [31], Fan and Wang proposed a logging mechanism to ensure the completeness of logs. By applying steganography to log forensics, the intrusion altered records can be kept, ensuring the completeness of the logs. In [32], Farnaaz and Jabbar built a model using a random forest as a classifier for intrusion detection. The proposed method is applied to the detection of four types of attacks, DoS, probe, User to Root (U2R), and Remote to Local (R2L) and evaluated using NSL-KDD dataset. Experimental results show that the proposed method using the random forest classifier performs better than other traditional classifiers. In [33], Shone et al. proposed a deep learning based intrusion detection method to handle concerns such as increasing levels of required human interaction and poor detection accuracy. Specifically, the proposed method is a combination of deep and shallowing learning. The deep learning refers to a non-symmetric deep auto-encoder and shallow learning refers to a random forest. With the stacked algorithm, this method is capable of classifying a wide range of network traffic correctly.

5.3 Attack Graph-Based Techniques

An attack graph can be used to identify all possible paths of attacks launched by an intruder in a network through analysis of hosts, networks and other security devices during the attack. Moreover, the attack graph can help the network administrator in defence before attacks happen by visualizing attack paths and determining the worst attack paths situation [30]. Therefore, in contrast to other network forensic techniques, attack graphs can be used a preventative measure to detect problems before they arise. In [34], Liu et al. proposed capability enhanced attack graphs by integrating anti-forensic techniques with attack graphs so as to recreate attack scenarios in the presence of anti-forensics tools/techniques. In [35], Albanese et al. introduced a new framework to analyse alerts generated in the network in real time and measure the impact of attacks. In this paper, the notation generalized dependency graph is first introduced. The generalized dependency graph can tell how network components depend on each other. In [36], Musa et al. presented an effective model with attack graphs to depict the devices and data flow. It can efficiently identify the weakest nodes along with the concerned vulnerability's origin. Moreover, with the use of risk and Common Vulnerability Scoring System (CVSS) based scores as evaluation criteria in the proposed approach, the time taken to identify the attack paths is greatly reduced.

5.4 Honeypots

A computer or computer system which is set up to mimic the possible targets of a cyberattack is called a honeypot. The honeypot is utilized to detect attacks or deflect/steer attacks from a legitimate target. It can also be used to obtain information about the operation of cybercriminals.[9] Therefore, the most important value of a honeypot is being detected, attacked or compromised [37]. A honeypot works by having the attacker interact with it. It is worth noting that a honeypot is not a solution in itself and therefore a honeypot does not address a specific problem. However, the honeypot is a highly flexible and effective tool with multiple applications in preventing attacks, detecting unauthorised activities and gathering intelligence on attackers [37]. Honeypots can be classified into low-interaction honeypots and high-interaction honeypots based on the level of interaction that they can provide. Also, honeypots can be categorised into production and research honeypots, real and virtual honeypots. Systems that control critical infrastructure or Supervisory Control and Data Acquisition (SCADA) systems are often a target of hackers. Jicha et al. in [38] conducted an analysis of one SCADA honeypot so as to determine the viability of the SCADA honeypot. In [39], different honeypots and their applications to the smart grid are investigated and compared. In [40], Pogliani et al. developed and deployed a honeypot that can simulate a complex distributed control system (DCS) as used in the industrial sector. In this way, they can record the attempts of attacks targeting the DCS and gain useful knowledge for research purposes.

5.5   Privacy-Preserving Data Analytics

Network data are generated when people interact with connected devices, viz., computers and networks (e.g., corporate wireless networks). Network security data can contain identifiers (e.g., usernames) that are closely related to people in the real world [41]. As privacy awareness continues to increase, both academia and industry have been actively developing viable solutions to meet the ever-growing privacy protection requirements [42]. In [43], Afanasyev et al. presented a cryptographic signature method at the packet level and permits authorised parties to assess any packet and physical machine that send the packet. This packet signature-based method does not leak identification information to the attacker. In [41], DeYoung developed a method for supporting the proper publication of network security data. In this research, a database that contains the network security data of a contextualised network and some other information (e.g., users' interaction with the network points) is built, but the database has no identifiability problem. In [44], Keshk et al. proposed a privacy-preserving anomaly detection framework, named PPAD-CPS, to protect confidential information and discover malicious data patterns/observations in network traffic. The proposed framework consists of two modules, the data preprocessing module and anomaly detection module. The experimental results on two publicly available datasets demonstrate that the proposed framework is effective and outperforms many existing anomaly detection techniques.

---

[9] https://us.norton.com/internetsecurity-iot-what-is-a-honeypot.html.

6   Research Gaps in Network Forensics

6.1   High Speed and Large Volume of Network Data

In high-speed networks, to record all packets without loss is a challenging task. Millions of packets are transmitted through such a network every second and network devices for network forensics are supposed to be able to capture, record and analyse network data as they flow through the network. However, network traffic data usually cannot be recorded completely by devices. This will result in incomplete logs of network traffic, making the reconstruction of suspicious attacks and the identification of attack sources more difficult [30]. Moreover, in network forensics, large volumes of data are collected from various devices, making evidence retrieval from a network problematic, as compared to simple, single-protocol, homogenous-device networks. For example, data should be captured on devices with sufficient storage space, but the problem of assuring intergrity for large files is often underrated. Therefore, on-the-fly processing of data becomes more important, as it avoids the requirement of large storage in traditional "store and process" methods [45].

6.2   Network Data Privacy

Data collected/generated by IoT devices and transmitted through one or more networks may contain sensitive personal information. For instance, data collected by smart wearable devices include a user's health information as well as potentially that user's identity. Therefore, network data should be carefully managed to avoid the leakage of user privacy. Perera et al. [46] suggest that existing technologies are insufficient to support a privacy-guaranteed data management lifecycle; new technologies and/or comprehensive anonymisation frameworks are needed in the IoT that can anonymize data communication paths. Ali et al. [47] proposed a network architecture which can provide IoT data privacy via two techniques, the blockchain and Interplanetary file system (IPFS). In the proposed architecture, network access control and accountability for data owners/users are enforced by the blockchain smart contracts. In addition to IoT applications, there are other applications for network data analysis, such as big data in power systems or social media. Therefore, it is necessary to develop privacy-preserving methods for specific applications under different threat models. Moreover, data analytic techniques for encrypted data based on deep learning models are an emerging technology that uses encryption algorithms without degrading system performance [42].

6.3   Artificial Intelligence (AI) in Network Forensics

The major objective of network forensics is to recognize all possible security breaches and create these features in a detection and prevention mechanism to prevent further damage. Artificial intelligence (AI) and its sub-field, machine learning (ML) offer a powerful set of tools to achieve this objective through its capability of feature classification. For instance, Mukkamal and Sung [48] proposed two AI techniques, artificial neural networks (ANN) and support vector machines (SVM), for offline intrusion detection. Moreover, in this work, ranking the features as an important step of feature selection based on importance is studied. Using AI techniques enables immediate effects to be achieved, however, there are also several limitations of AI in network forensics. For example, the performance of AI-based systems heavily relies on training data sets. In fact, in [49] it is noted that the choice of real-world data sets and feature selection are more important than the AI algorithm itself. Nonetheless, research into AI algorithms continues [50], and is important. If a malicious pattern is reported for the first time, then AI-based schemes are unable to handle it. Moreover, AI-based systems usually require a huge amount of data to train the model so as to be accurate and powerful in classification, however, in many cases, it could be hard to provide the system with a large amount of data for training from the point of view of a network forensic investigator [51]. Therefore, more work needs to be done to allow network forensics to take full advantage of the benefits AI provides [52].

6.4   Improving Honeypot Development

Honeypot as a decoy-based intrusion protection technology has been used in various aspects of security, such as detection, information gathering and prevention [53]. According to [54], honeypots can be divided into two categories, namely, production honeypots and research honeypots. Specifically, production honeypots are used to collect data on adversaries trying to gain access to an organisation's infrastructure. Research honeypots, on the other hand, are used to collect data for analysing the techniques and schemes used by attackers to gain unauthorised access. In recent years, IoT is drawing more attention from both society and industry. It is predicted that tens of billions of IoT devices will be deployed by 2020 [55]. However, until now only limited efforts (e.g., [55–57]) have been made in building a comprehensive honeypot that emulates an IoT platform. In the future, more efforts should be made in improving honeypots to be resilient against anti-forensics methods, to handle huge quantities of traffic data, and to increase the range of imitated IoT devices [58]. According to [59], the following factors impede the development of honeypots: (1) It is hard to tune or configure honeypots. A honeypot is configured to attract and trap aiming targets. However, if the honeypot is wrongly configured/tuned, it will not only fail to attract its prey but could be detected and hijacked by adversaries. (2) The legal and ethical issues associated with the use of honeypots are taken seriously in the field of network management. The most fundamental problem is that there is no consensus in the cyber community about what activities are deemed acceptable or unacceptable. Considering the above inhibiting factors, more work is required to improve the development of honeypots.

6.5 E-Discovery

As noted above, network environments are complex, multi-protocol and heterogeneous. The evolving nature of the threat landscape, coupled with the desire for increased connectivity, has meant that hybrid solutions are likely. Such solutions involve machine-speed responses to events, with experienced analysts performing post-incident triage. Notwithstanding the comments above, efforts have been made in various areas (e.g., intrusion detection) to use machine learning as a "first responder". This strategy has not been as effective as claimed for a variety of reasons. The prime reason is that a machine learning algorithm tends to require significant tuning and interpretation. As a result, it does not reduce the need for a trained workforce; it actually exacerbates the problem.

Nonetheless, automated search and retrieval of data (called e-Discovery) is a valid and legitimate research area, especially for forensic purposes [60]. Electronic data has properties that make it particularly amenable to e-Discovery. Data can come from several disparate sources (e.g., emails, tablets or smartphones); in different formats (documents, spreadsheets and pictures); and in different locations (a local computer, a corporate server and a cloud service). Finally, it is difficult to completely destroy data.

7 Explainable Artificial Intelligence for Network Forensics

With machine learning being a core technology, artificial intelligence has achieved significant success in tackling incrementally sophisticated computing tasks. AI is also a key aspect of the future development of network forensics. However, for black-box techniques (e.g., deep neural networks) [61], interpretability has become an issue. In the case of law enforcement, it is essential that decisions made by AI-based tools and those made with their assistance be justifiable and explainable to humans [62]. Explainable artificial intelligence (XAI) attempts to solve the black box problem, where machine learning algorithms often reach decisions using methods that cannot be specifically explained. A truly interpretable AI with a rationalisable black box can offer subjective forensic approaches more confidence by providing both a transparent process and a sound logical reasoning. In a truly interpretable AI system, not only can a user see, but he/she is also able to examine and understand how the inputs are mapped mathematically to the outputs [62].

Interpretability can be offered by means that include, but not limited to, linear/logistic regression, generalised additive models and decision rules [63]. To obtain a balanced trade-off between interpretability and predictive accuracy of AI algorithms, techniques such as simplification, feature correlation estimation, or visualisation can be used to simulate black-box models of AI algorithms, with the aim of converting an opaque model (the black box) into a glass box. Some examples of such techniques include locally interpretable model interpretation [64] and SHapley Additive exPlanations [65], which compute the contribution of each feature to the prediction based on three desirable properties, namely, local accuracy, missingness and consistency. These are referred to as XAI, aiming to produce a pool of ML techniques to attain more interpretable, understandable and trustworthy models with no significant loss in prediction accuracy [66].

According to [62], the implementation of AI in the field of digital/network forensics is currently still black-box in nature, requiring validation by digital/network forensic examiners and thus unlikely to be explainable in court. The admissibility in court of digital/network analysis conducted by XAI is currently uncertain, as it would require an inspection of existing legislation (e.g., evidence law). However, if XAI is not seen as a substitute for a digital/network forensic examiner, but instead as a robust tool to assist in investigations, then XAI can be utilised effectively and legitimately to augment the digital/network forensic field in the future.

8 Conclusion

Network forensics focuses on capturing and analyzing network traffic so as to detect abnormal traffic and identify intrusions, which is an important research topic in the era of AI that offers both academics and practitioners opportunities to solve challenging problems brought about by the complexity of real-world networks and the almost ubiquitous use of the Internet for commerce. This chapter briefly introduced the definition of network forensics, some common types of attacks targeting networks and existing network forensic tools. Moreover, this chapter described the current development of network forensics techniques and identified some specific research gaps in the present study of network forensics, thereby shedding light on future research directions.

Acknowledgements The work has been supported by the Cyber Security Research Centre Limited whose activities are partially funded by the Australian Government Cooperative Research Centres (CRC) Program.

References

1. S. Garfinkel, Network forensics: tapping the internet. IEEE Internet Comput. **6**, 60–66 (2002)
2. G. Palmer, A road map for digital forensic research. in *First digital forensic research workshop*, Utica, New York, pp. 27–30 (2001)
3. W. Yang, M. N. Johnstone, L. F. Sikos, S. Wang, Security and forensics in the internet of things: research advances and challenges. in *Workshop on Emerging Technologies for Security in IoT (ETSecIoT)*, Sydney, Australia (2020)
4. R. Mercuri, Courtroom considerations in digital image forensics. Digital image forensics, pp. 313–325, Springer (2013)
5. M. Meyers, M. Rogers, Digital forensics: meeting the challenges of scientific evidence. in *IFIP International Conference on Digital Forensics*, pp. 43–50, Springer (2005)
6. J. R. Vacca, Computer and information security handbook. Newnes (2012)
7. C. C. Wood, W. W. Banks, S. B. Guarro, A. A. Garcia, V. E. Hampel, H. P. Sartorio, Computer security: a comprehensive controls checklist. Wiley-Interscience (1987)
8. S. L. Garfinkel, Digital forensics research: the next 10 years. digital investigation **7**, S64-S73 (2010)
9. Z.A. Baig, P. Szewczyk, C. Valli, P. Rabadia, P. Hannay, M. Chernyshev, M. Johnstone, P. Kerai, A. Ibrahim, K. Sansurooah, Future challenges for smart cities: cyber-security and digital forensics. Digit. Investig. **22**, 3–13 (2017)
10. N. Kumari, A. Mohapatra, An insight into digital forensics branches and tools. in *2016 International Conference on Computational Techniques in Information and Communication Technologies (ICCTICT)*, pp. 243–250, IEEE (2016)
11. N.M. Karie, H.S. Venter, Toward a general ontology for digital forensic disciplines. J. Forensic Sci. **59**, 1231–1241 (2014)
12. W.G. Kruse II, J. G. Heiser, *Computer Forensics: Incident Response Essentials*. Pearson Education (2001)
13. R. Hunt, S. Zeadally, Network forensics: an analysis of techniques, tools, and trends. Computer **45**, 36–43 (2012)
14. R. Ahmed, R. V. Dharaskar, Mobile forensics: an overview, tools, future trends and challenges from law enforcement perspective. in *6th International Conference on E-Governance, ICEG, Emerging Technologies in E-Government, M-Government*, pp. 312–323 (2008)
15. M.A. Guimaraes, R. Austin, H. Said, Database forensics. in *2010 Information Security Curriculum Development Conference*, pp. 62–65, ACM, (2010)
16. M. Burdach, *Physical memory forensics* (Black Hat, USA, 2006)
17. F. Armknecht, A. Dewald, Privacy-preserving email forensics. Digit. Investig. **14**, S127–S136 (2015)
18. A. Powell, C. Haynes, Social media data in digital forensics investigations. in *Digital Forensic Education* (Springer, 2020), pp. 281–303
19. S. Simou, C. Kalloniatis, E. Kavakli, S. Gritzalis, Cloud forensics solutions: a review. in *International Conference on Advanced Information Systems Engineering*, pp. 299–309, Springer (2014)
20. S. Alabdulsalam, K. Schaefer, T. Kechadi, N. -A. Le-Khac, Internet of things forensics–challenges and a case study. in *IFIP International Conference on Digital Forensics*, pp. 35–48, Springer (2018)
21. S. Datt, *Learning Network Forensics* (Packt Publishing Ltd 2016)
22. E. S. Pilli, R. C. Joshi, R. Niyogi, Network forensic frameworks: survey and research challenges. Digit. Investig. 7, 14–27 (2010)
23. Top 8 network attack types (2017) https://www.calyptix.com/top-threats/top-8-network-attacks-type-2017
24. A. Woodiss-Field, M. N. Johnstone, BotNets in the internet of things: the next wave. in *Conference Proceedings of the 18th Australian Cyber Warfare Conference 2019 (CWAR 2019)*, pp. 41–45 (2020)

25. N. E. Hastings, P. A. McLean, TCP/IP spoofing fundamentals. in *Conference Proceedings of the 1996 IEEE Fifteenth Annual International Phoenix Conference on Computers and Communications*, pp. 218–224. IEEE (1996)
26. X.-J. Wang, X.-y. Wang, Topology-assisted deterministic packet marking for IP traceback. J. China Univ. Posts Telecommun. **17**, 116–121 (2010)
27. Y. Fen, Z. Hui, C. Shuang-shuang, Y. Xin-chun, A lightweight IP traceback scheme depending on TTL. Proced. Eng. **29**, 1932–1937 (2012)
28. E. Jeong, B. Lee, An IP traceback protocol using a compressed hash table, a Sinkhole router and data mining based on network forensics against network attacks. Futur. Gener. Comput. Syst. **33**, 42–52 (2014)
29. J. J. Kang, L. F. Sikos, W. Yang, Reducing the attack surface of edge computing IoT networks via hybrid routing using dedicated nodes. Secur. Edge Comput. Appl. Tech. Chall. 97–111 (2021)
30. S. Khan, A. Gani, A.W.A. Wahab, M. Shiraz, I. Ahmad, Network forensics: review, taxonomy, and open challenges. J. Netw. Comput. Appl. **66**, 214–235 (2016)
31. Y. -T. Fan, S. -J. Wang, Intrusion investigations with data-hiding for computer log-file forensics. in *2010 5th International Conference on Future Information Technology*, pp. 1–6, IEEE (2010)
32. N. Farnaaz, M. Jabbar, Random forest modeling for network intrusion detection system. Proced. Comput. Sci. **89**, 213–217 (2016)
33. N. Shone, T.N. Ngoc, V.D. Phai, Q. Shi, A deep learning approach to network intrusion detection. IEEE Trans. Emerg. Top. Comput. Intell. **2**, 41–50 (2018)
34. C. Liu, A. Singhal, D. Wijesekera, Using attack graphs in forensic examinations. in *2012 Seventh International Conference on Availability, Reliability and Security*, pp. 596–603, IEEE (2012)
35. M. Albanese, S. Jajodia, A. Pugliese, V. Subrahmanian, Scalable analysis of attack scenarios. in *European Symposium on Research in Computer Security*, pp. 416–433, Springer (2011)
36. T. Musa, K. C. Yeo, S. Azam, B. Shanmugam, A. Karim, F. De Boer, F. N. Nur, F. Faisal, Analysis of complex networks for security issues using attack graph. in *2019 International Conference on Computer Communication and Informatics (ICCCI)*, pp. 1–6, IEEE (2019)
37. V. Maheswari, P. Sankaranarayanan, Honeypots: deployment and data forensic analysis. in *International Conference on Computational Intelligence and Multimedia Applications (ICCIMA 2007)*, pp. 129–131, IEEE
38. A. Jicha, M. Patton, H. Chen, SCADA honeypots: an in-depth analysis of Conpot. in *2016 IEEE conference on intelligence and security informatics (ISI)*, pp. 196–198, IEEE (2016)
39. C. Dalamagkas, P. Sarigiannidis, D. Ioannidis, E. Iturbe, O. Nikolis, F. Ramos, E. Rios, A. Sarigiannidis, D. Tzovaras, A survey on honeypots, honeynets and their applications on smart grid. in *2019 IEEE Conference on Network Softwarization (NetSoft)*, pp. 93–100, IEEE (2019)
40. M. Pogliani, M. Barlocco, M. Pesatori, S. Zanero, Poster: using honeypots to understand attacks to industrial control systems (2019)
41. M.E. DeYoung, *Privacy Preserving Network Security Data Analytics* (Virginia Tech, 2018)
42. H.-Y. Tran, J. Hu, Privacy-preserving big data analytics a comprehensive survey. J. Parallel Distrib. Comput. **134**, 207–218 (2019)
43. M. Afanasyev, T. Kohno, J. Ma, N. Murphy, S. Savage, A.C. Snoeren, G.M. Voelker, Privacy-preserving network forensics. Commun. ACM **54**, 78–87 (2011)
44. M. Keshk, E. Sitnikova, N. Moustafa, J. Hu, I. Khalil, An integrated framework for privacy-preserving based anomaly detection for cyber-physical systems. IEEE Trans. Sustain. Comput. (2019)
45. I. Yaqoob, I.A.T. Hashem, A. Ahmed, S.A. Kazmi, C.S. Hong, Internet of things forensics: recent advances, taxonomy, requirements, and open challenges. Futur. Gener. Comput. Syst. **92**, 265–275 (2019)
46. C. Perera, R. Ranjan, L. Wang, S.U. Khan, A.Y. Zomaya, Big data privacy in the internet of things era. IT Prof. **17**, 32–39 (2015)
47. M. S. Ali, K. Dolui, F. Antonelli, IoT data privacy via blockchains and IPFS. in *Proceedings of the Seventh International Conference on the Internet of Things*, pp. 1–7 (2017)

48. S. Mukkamala, A.H. Sung, Identifying significant features for network forensic analysis using artificial intelligent techniques. Int. J. Digital Evid **1**, 1–17 (2003)

49. M. N. Johnstone, M. Peacock, Seven pitfalls of using data science in cybersecurity, in *Data Science in Cybersecurity and Cyberthreat Intelligence*, ed. by Sikos, L.F., Choo, K.-K., R. (Springer, Cham, Switzerland, 2020)

50. W. Yang, S. Wang, M. Johnstone, A comparative study of ML-ELM and DNN for intrusion detection. in *2021 Australasian Computer Science Week Multiconference*, pp. 1–7 (2021)

51. P. H. Rughani, Artificial intelligence based digital forensics framework. Int. J. Adv. Res. Comput. Sci. **8** (2017)

52. F. Mitchell, The use of Artificial Intelligence in digital forensics: an introduction. Digital Evid. Elec. Signat. L. Rev. **7**, 35 (2010)

53. I. Mokube, M. Adams, Honeypots: concepts, approaches, and challenges. In *Proceedings of the 45th Annual Southeast Regional Conference*, pp. 321–326, ACM (2007)

54. P. N. Rabadia, Extraction of patterns in selected network traffic for a precise and efficient intrusion detection approach. vol. PhD. Edith Cowan University (2018)

55. M. Wang, J. Santillan, F. Kuipers, ThingPot: an interactive Internet-of-Things honeypot. arXiv: 1807.04114 (2018)

56. Y.M.P. Pa, S. Suzuki, K. Yoshioka, T. Matsumoto, T. Kasama, C. Rossow, IoTPOT: analysing the rise of IoT compromises. *In 9th {USENIX} Workshop on Offensive Technologies ({WOOT} 15* (2015)

57. J. D. Guarnizo, A. Tambe, S. S. Bhunia, M. Ochoa, N. O. Tippenhauer, A. Shabtai, Y. Elovici, Siphon: towards scalable high-interaction physical honeypots. in *Proceedings of the 3rd ACM Workshop on Cyber-Physical System Security*, pp. 57–68, ACM (2017)

58. N. Koroniotis, N. Moustafa, E. Sitnikova, Forensics and deep learning mechanisms for botnets in internet of things: a survey of challenges and solutions. IEEE Access **7**, 61764–61785 (2019)

59. M.L. Bringer, C.A. Chelmecki, H. Fujinoki, A survey: recent advances and future trends in honeypot research. Int. J. Comput. Netw. Inf. Secur. **4**, 63 (2012)

60. D.W. Oard, W. Webber, Information retrieval for e-discovery. Inf. Retr. **7**, 99–237 (2013)

61. L. Kelly, S. Sachan, L. Ni, F Almaghrabi, R. Allmendinger, Y. -W. Chen, Explainable artificial intelligence for digital forensics: opportunities, challenges and a drug testing case study. Digital Forens. Sci. IntechOpen (2020)

62. S. W. Hall, A. Sakzad, K. K. R. Choo, Explainable artificial intelligence for digital forensics. Wiley Interdiscip. Rev. Forens. Sci. e1434 (2021)

63. C. Molnar, Interpretable machine learning. Lulu. com (2020)

64. M. T. Ribeiro, S. Singh, C. Guestrin, Why should i trust you?" Explaining the predictions of any classifier. in *Proceedings of the 22nd ACM SIGKDD International Conference on Knowledge Discovery and Data Mining*, pp. 1135–1144 (2016)

65. S. M. Lundberg, S. -I. Lee, A unified approach to interpreting model predictions. in *Proceedings of the 31st International Conference on Neural Information Processing Systems*, pp. 4768–4777 (2017)

66. D. Gunning, M. Stefik, J. Choi, T. Miller, S. Stumpf, G.-Z. Yang, XAI—Explainable artificial intelligence. Sci. Robot. 4, eaay7120 (2019)

Chapter 9
Obfuscation-Based Mechanisms in Location-Based Privacy Protection

Basia E. Wozny, Lei Pan, and Adnan Anwar

Abstract Protection of personal data is vital to individuals. With the prevalence of cyber-crime, it is important to leverage the power of artificial intelligence (AI) to protect the user. The more someone else knows about a person and their whereabouts, the more power they can have over that person. This power takes the form of influence over their decisions, perhaps reveals to others their behaviours, where they live, and subsequently can affect their reputation or even safety. This review will provide an exploration into three methods of obscuring a person's geographical location when using location-based services. It will explain to the reader how artificial intelligence can be used in their favour. Unfortunately, there is no perfect protection of privacy because location-based services (LBS) rely on the device's accurate location to provide services, such as restaurant recommendations or games like Pokemon. But we should at least aim for a favourable balance between total privacy and the exposure of information on an individual. Three obfuscation-based mechanisms were compared on their data privacy efficacy, data utility, and ease of implementation. These mechanisms are: location generalisation (cloaking), location perturbation (differential privacy), and location spoofing (using dummy locations). Gowalla is a location-based social networking website where users share their locations by checking-in. A subset of the Gowalla dataset was used to evaluate these three mechanisms. In this work, three obfuscation methodologies for location-based services to protect the user's location and privacy were evaluated. For the Gowalla dataset subset in London, the differential privacy using logistic regression with an epsilon of 0.8 seemed to have the best performance in terms of privacy, utility and overhead. k-Anonymity came second in its reasonable performance across all three metrics. t-Closeness obfuscation performed better in privacy, but worse than l-diversity in utility. Finally, these were followed by the application of dummy based locations. This method was the simplest but was the least refined in its ability to hide user datapoints and the most vulnerable to malicious attacks. The differential privacy

B. E. Wozny · L. Pan · A. Anwar (✉)
Centre for Cyber Security Research and Innovation (CSRI), School of IT, Deakin University, Geelong 3216, Australia
e-mail: adnan.anwar@deakin.edu.au

© The Author(s), under exclusive license to Springer Nature Switzerland AG 2022
M. Ahmed et al. (eds.), *Explainable Artificial Intelligence for Cyber Security*,
Studies in Computational Intelligence 1025,
https://doi.org/10.1007/978-3-030-96630-0_9

using k-means was considered separately, since it is an unsupervised artificial intelligence learning algorithm as opposed to the others. Differential privacy clustering showed the ease of being able to create varying clusters each time a query is run, which could be highly effective in hiding user device locations.

Keywords Location privacy · k-Anonymity · Gowalla dataset · Obfuscation

1 Introduction

Location-based services are increasingly popular in providing users with useful information about their surroundings given their whereabouts, for example: eatery recommendations. These services depend on data utility (usability of the data) which is paramount in providing relevant information. However, the higher the data utility, the lower the user privacy. To increase user privacy, we can employ mechanisms such as cloaking, 'data noise', and data aggregation to hide the user's data points in amongst other data points.

This review investigates the deployment of these mechanisms via algorithms of k-anonymity, k-means/logistic regression differential privacy, and using 'fakers' to spoof would-be attackers that could work out the location of a user.

These three mechanisms are just some of many other ways to protect privacy. Hongbo, et al. [1], describes additional privacy mechanisms not addressed in this review such as:

- Privacy policies which involve agreements between parties as to the privacy they will apply to the data,
- Cryptography including space transformation, secure multiparty computation (SMC), and private information retrieval (PIR)
- Co-operation and caching which is the action of reducing the number of connections with location-based servers as much as possible and at the same time reducing user exposure.

Alternatively, users can also potentially choose to share or not to share their location data, which is a factor in holistic data privacy.

The first mechanism, k-anonymity, is based on the idea that there is a trusted Central Authority that takes the location from the user and cloaks it or obfuscates the user's location before passing it to the service provider. On its return from the service provider, the cloaking is removed, and the result is presented to the user [2]. Cloaking involves aggregating or grouping the features so that individuals cannot be distinguished by those features. Deutch et al. [3] looked at optimising the trade-off between privacy and utility in data provenance. Data provenance is equivalent to data lineage which provides the where, what and how of the datum [4]. This paper looked at the tension between releasing too much information and hiding information. Privacy-wise, k-anonymity attackers can use the probability of the location information to identify where the user is located. Privacy is based on how many queries match the obfuscated lineage. Fei et al. [2] discovered a degrading quality of service with k-anonymity because the location information is coarse-grained, so data utility is low. Deutch et al. [3] attempted to find the balance between privacy measures and using the data effectively. They quantified the loss of information using entropy. When k-anonymity was used with k = 23, 90% of the runs of the algorithm returned at least twice the level of privacy but, loss of information increased as privacy increased and therefore resulted in a decrease in utility. Slijepčević, et al. [5] found that high accuracy is achieved if the generalised dataset is strongly correlated with the target variable. However, if suppression is used, it is increasingly difficult to get the accuracy required. Suppression is advisable as it allows for flexibility in the generalisation of the data. Deutch et al. [3] found that k-anonymity is highly scalable even for large k (privacy threshold).

Another emerging technique, known as differential privacy, describes the addition of Laplacian noise to a dataset to obscure the true data points. k-Means clustering can be used in conjunction with differential privacy methodology to achieve this. An inherent risk with k-means clustering is that when it comes to outliers that do not belong to a cluster, or are in a cluster by themselves, they can be identified [6]. Yu et al. [6] stated that compared to k-anonymity, differential privacy does not require continual improvements in response to new attacks. It avoids attacks based on background knowledge (e.g., knowing the geographical terrain), applying privacy by adding random noise to the results. They also proposed an outlier-eliminated differential-privacy which can combat the identification of loners in the dataset, where the outlier was hidden by the densities of the r-nearest-neighbour. This also allegedly combats the loss of information where the clustering results they achieved were close to that of the original data. Regarding ease of implementation, Lin [7] describes a method to deal with the computational volume by outsourcing with a random linear transformation. The data is sent to the service provider, and they solve the kernel k-means from the encrypted data. This provides large-scale clustering performance with little overhead to the data owner.

The third methodology, unattractively called 'spoofing', a word commonly associated with attackers, is in this context the method by which one spams an adversary with fake locations. Several dummy users are generated and together with the user's actual location are sent to the location-based service provider. The user gets their results by filtering out the unnecessary results. The key privacy issue with this methodology is that outliers can be found with additional background information [1]. Generated dummy users need to be viable and humanlike. For example, it would not make sense for a user to be moving at unrealistic speeds [6]. Hara et al. [8] conducted tests with humans to see if they could identify the real users from the dummies. The dummies they generated considered the geography of the user, and interestingly, they were able to hide the users well, but not where real-life issues are concerned. This is confirmed by Du et al. [9], "side-information" is available to attackers and can reveal the user. The spoofing method has advantages over the others in that the recommender results are generally accurate [1]. In the Du et al. [9] study, entropy was not lost using their Enhanced Dummy-Based Location Privacy. However, this method had issues with being an intense process, needing to send many bogus signals, and on the receiving end, the location-based servers have to deal with the spoofing impact [1]. Furthermore, there are also issues implementing ideal parameters, which can be difficult when considering real-life factors [10]. While security has been a key focus across multiple critical infrastructure domains [11–13], in this work we have focused mainly on the privacy aspects.

In this review the three obfuscation methods will be evaluated on their privacy capability, utility, and service overhead. The review is organized as follows:

- Performance metrics for algorithm comparison.
- Dataset preparation and exploration.
- Application of k-anonymity using the Mondrian algorithm by Kristen et al. [14] and applying it with two algorithm implementations: l-diversity and t-closeness.
- Application of the IBM differential privacy algorithms [15, 16] using both k-means clustering and logistic regression.
- Spoofing: A rudimentary application of randomly dummy-based locations.
- Comparative results.

2   Performance Metrics and Their Measurement

Three main performance metrics allow the comparison of the algorithms.

Privacy Metric—Matching /reidentification of data: Each algorithm in this review has its performance measure of privacy. K-anonymity primarily uses entropy to quantify information loss [3], while in differential privacy using k-means, disclosure risk is used to identify before and after adding noise to a cluster [16, 17]. The dummy location success can also be measured using entropy.

Utility Metric—Ability to use the data for meaningful purposes e.g., research and Location Based Services (LBS) functions: The performance metric for utility is whether the service is degraded, or in algorithm terms, whether the performance of predictions or clustering is achieved at a level close to the obscured data.

Overhead Metric—Work done by servers and devices: There are a few overhead metrics: execution time of the algorithm, bandwidth and energy utilisation, and storage. Impacts on both sides of the fence need to be considered, from the user side and the LBS side.

3   Implementation and Performance of the Privacy Methods

3.1   Dataset Used

The Gowalla dataset [18] was utilised in this implementation of the above researched methodologies. Since it is a rather large dataset, a subset was taken to include data points in a rectangular area in London. The area includes Bromley, Sydenham,

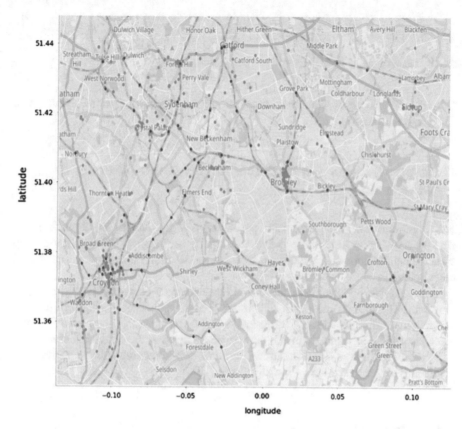

Fig. 1 User locations as green points with train stations as red points

Sidcup, St Mary Cray, and Croydon. Train stations (red points) were added as they would be used to suggest stations in the area around the users (green points). See Fig. 1.

Python was used to implement the methodologies on this dataset. The code used can be accessed here: https://github.com/bwozny/Location-based-Privacy-Protection.

3.2 Data Preparation/Exploration

The following is a summary of steps performed on the dataset:

1. Split the date into date and time components: year, month, weekday and hour.
2. Add station location for user recommendations (not utilised fully).
3. Retrieve from Nominatim (a geocoder) the geolocation of the user locations.
4. Dropping features depending on the algorithm used.

user	check_in_time	latitude	longitude	location_id	...	checkin_month	checkin_day	checkin_hour	checkin_weekday	Address
14009	72	2010-04-18T07:04:21Z	51.442041	-0.00445	808780	...	4	18	7	Sunday	Sandhurst Road, Catford South, London Borough ...

Fig. 2 Data preparation—Sample row

5. For all algorithms, a privacy baseline is established on the non-private data (un-obfuscated) (Fig. 2).

4 Applying k-Anonymity Methodology with l-Diversity & t-Closeness

The algorithm splits the data into partitions using a Mondrian algorithm and aims to statistically hide the sensitive features by aggregation [10] using a greedy approximation algorithm with $k = 3$ and $k = 10$ (common number of nearest neighbours).

The sensitive feature being the user id and the key features, the latitude and longitude. The address and location id would need to be removed during LBS process as this would directly identify the user. Additional features of the check-in hour, weekday and month could be used, but for simplicity they are not considered.

The way this works is that the individual data points become indistinguishable in a generalised group created by the algorithm [18].

4.1 Privacy Metric

The application of k-anonymity impacts the performance of a dataset that has been treated with l-diversity and t-closeness, as shown in Table 1. A baseline performance is established by running logistic regression on the raw data subset from Gowalla.

Table 1 Results for All 3 datasets using logistic regression

k-nearest neighbours	Implementation	Non-private data (Baseline) (%)	Mondrian k-Anonymity (%)	Mondrian l-Diversity (%)	Mondrian t-Closeness (%)
$k = 3$	Algorithm Performance	66.44	26.08	16.74	16.67
	Privacy metric	33.56	73.92	83.26	83.33
$k = 10$	Algorithm Performance	66.44	14.76	14.04	10.80
	Privacy metric	33.56	85.24	85.96	89.20

The higher the performance value above, the lower the privacy. The ability of the algorithm to "predict" the data points also further decreased with an increase in k nearest neighbours for aggregation to each datapoint (k = 3 → k = 10).

l-Diversity considers the diversity of sensitive values in the group, but does not take into.

account the semantic closeness of the values. This could leave the user open to a similarity attack. If there are no other data points similar, then the user is exposed, as in Fig. 4 in the utility section below.

t-Closeness is where the distance between the distribution of a sensitive attribute and the distribution of the attribute in the dataset is more than a threshold t. In this review the Kolmogorov–Smirnov distance is used [19]. However, there are many different types of distance measures that could be used such as Earth Mover Distance, Equal Distance, Ordered Distance, Hierarchical Distance and so on [20].

4.2 Additional Parameters

t-Closeness: the maximum allowed Kolmogorov–Smirnov threshold distance, t = 0.5

l-Diversity: the minimum required diversity of sensitive attribute values in the partition, l = 5. Increasing the Kolmogorov–Smirnov distance in the t-closeness partition or increasing the minimum number of sensitive values in the l-diverse partition decreases performance, and so, increases privacy in both algorithms.

4.3 Utility Metric

Below are 3 partitions of the dataset.

The first partition: the k-Anonymity Mondrian implementation, where the individuals are grouped into groups of at least k persons. The algorithm replaces the identifier attributes with aggregate values to hide individuals.

Figure 3 shows the baseline k-anonymity implementation with the number of nearest neighbours to each user location (k) is 3. The first k-anonymous partitioning (Mondrian) excludes some individual datapoints.

The second partition: the l-diversity implementation makes sure that each k-anonymous group contains at least "l" number of different values of the sensitive attribute (user location). Figure 4 shows how additional data points in violet are anonymised.

"l" Denotes minimum required diversity of sensitive attribute values in the partition, and for this review, it is 5. l-Diversity aggregations are of average utility. Users would get reasonable recommendations.

The third partition: The t-closeness criterion makes sure the statistical distribution of the attribute values in each k-anonymous group is "close" to the overall distribution

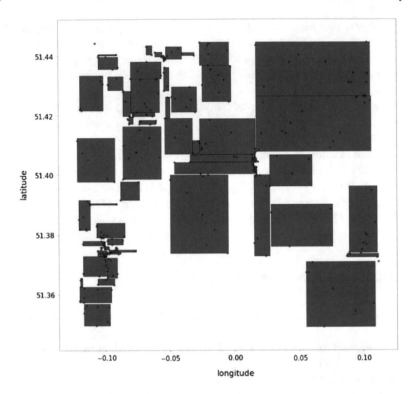

Fig. 3 K-Anonymity Implementation in maroon (k = 3)

of that attribute in the entire dataset [21]. Figure 5 shows that almost three suburbs of users have been merged into aggregate values (shown in violet overlay). It is important to note that the utility could be increased with the use of a distance measure between two probability distributions as described in Ninghui et al. [22].

The utility of the data decreases with the larger aggregations in Fig. 5. The user in the bottom right corner would get recommendations from somewhere in the middle of the map. The aggregations are too large to be able to give accurate recommendations.

4.4 Overhead Metric

In this review with the small dataset, the overhead is low, however for a larger dataset this could be a cumbersome algorithm for each server. This would be exacerbated by needing to remove the aggregation to provide an accurate service to the user or by the type of distance or diversity measures used.

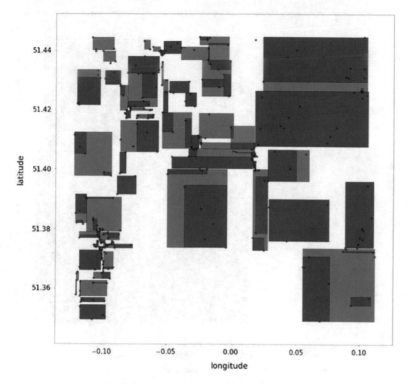

Fig. 4 1-Diversity Implementation overlay in violet ($k = 3, l = 5$)

5   Applying Differential Privacy Using K-Means and Logistic Regression Algorithms

The differential privacy mechanism applies Laplacian noise to the data to mask the real data points. Two algorithms were applied for differential privacy: the first was K-Means, and the second was Logistic Regression. The Differential privacy algorithms used were developed by IBM [15, 16].

It is worth noting here, that the k-means algorithm clusters together the user's latitude and longitude values. It is an unsupervised learning artificial intelligence method. This means that its performance cannot be compared to the other algorithms in this review that are using supervised learning via logistic regression. It is evaluated in this review for the purpose of presenting a larger picture of artificial intelligence and its use in LBS privacy protection.

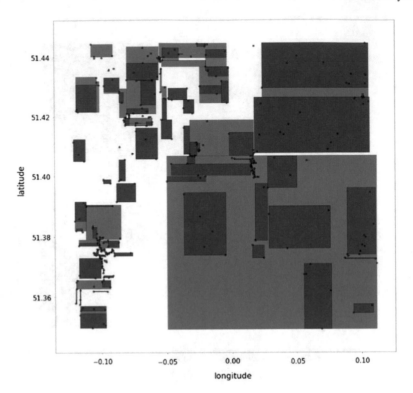

Fig. 5 T-Closeness Implementation overlay in violet (k = 3, Kolmogorov–Smirnov distance = 0.5)

5.1 Differential Privacy using K-Means

To determine how many clusters to start with, the dataset was plotted by city district, revealing 7 out of 10 potential clusters. The distribution of the users, however, was mostly concentrated in 3 of those as shown below in Fig. 6.

Plotting the inertia versus the number of clusters shows an "elbow criterion" where inertia slows, and there is no need for additional clusters. Figure 7 shows this occurs around 5 clusters. For this report, 5 clusters were used to apply differential privacy using the k-means algorithm.

5.2 Privacy Metric

After the application of epsilon of 0.1 the identification of the number of clusters needed becomes difficult, as seen with the erratic inertia in Fig. 8.

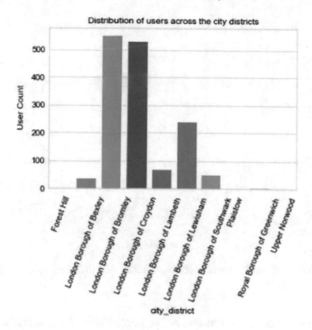

Fig. 6 Distribution of users across city districts

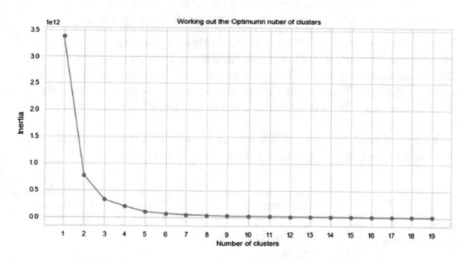

Fig. 7 Identification of Ideal number of clusters on non-private dataset

The privacy metric was measured by the application of clustering and comparing the silhouette score before and after application of differential privacy.

The silhouette score shows how well clusters are formed. Its value ranges from −1 to 1.

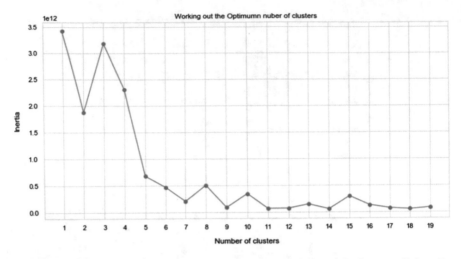

Fig. 8 Identification of Ideal number of clusters on data with differential privacy applied epsilon = 0.1

- Silhouette score of 1: The clusters are separated well from each other and can be easily distinguished.
- Silhouette score of 0: The distance between clusters is not significant.
- Silhouette score of −1: The clusters are assigned in the wrong way.

For the non-private data, with 5 clusters. The average silhouette score was: 0.631 whilst for the The results in Table 2 show that the non-private dataset had relatively clear clusters with an average silhouette score of 0.631. After application of Laplacian noise, the average silhouette score reduces to 0.408. It is important to note that this value changes each time the algorithm is run as the "noise" applied is different each time. See Figs. 9 and 10.

An adversary would not be able to pinpoint a user in the private dataset as each time the query is run, the user could belong to a different cluster.

Table 2 Silhouette scores before and after application of Differential Privacy by k-means clustering

Implementation	Non-Private baseline	Differential privacy with epsilon = 0.1
Average silhouette score	0.631	0.408
Privacy metric	Defined clusters	Undefined changing clusters

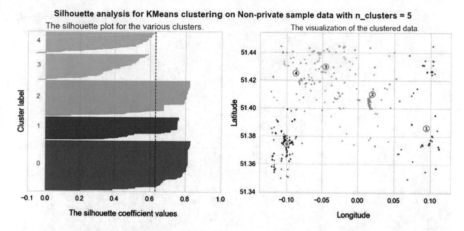

Fig. 9 Silhouette scores for Non-Private data

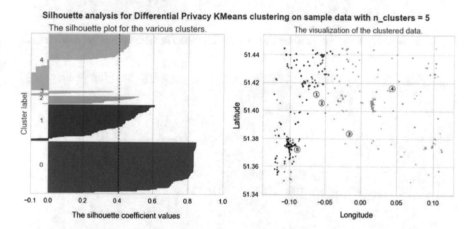

Fig. 10 Silhouette scores for Non-Private dataset with differential privacy applied epsilon $= 0.1$

5.3 Utility Metric

Provided the Laplacian noise is removed for returning results, this method would have high utility. To appreciate it visually, Fig. 11 shows clearly defined clusters, for the non-private dataset.

Figure 12 below, shows the shift of clusters after application of differential privacy with epsilon of 0.1.

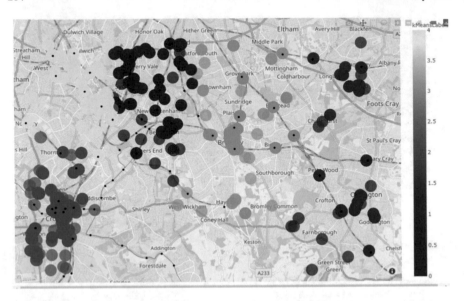

Fig. 11 Non-private data clusters based on Latitude & Longitude. Black points are train stations

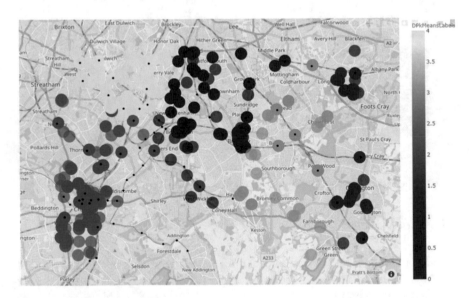

Fig. 12 DP Private data with epsilon 0.1 applied. Black points are train stations

5.4 Overhead Metric

As with the utility, the overhead is almost negligible for application of the Laplacian noise, Andrés, et al. [23], states that this is a straightforward implementation. Modern smart-phone devices would be computationally efficient and geo-indistinguishability are not excessive. There is an overhead cost that depends on the level of desired privacy and accuracy.

6 Differential Privacy with Logistic Regression

The logistic regression algorithm used here is obtained from the IBM's differential privacy library [16]. The IBM differential privacy logistic regression does not modify the distribution of the data as can be seen in Figs. 13 and 14.

6.1 Privacy Metric

The privacy metric here is measured by epsilon, the amount of Laplacian noise added to the data. The higher the epsilon, the closer it is to the baseline dataset [16]. The lower the epsilon value, the more noise is added to the data. Figure 15 shows the differentially private logistic regression accuracy versus epsilon for the user dataset. Table 3 shows the accuracy of the non-private dataset is approximately 73.74%, and the private dataset has a maximum accuracy of 0.69% (eps = 0.8).

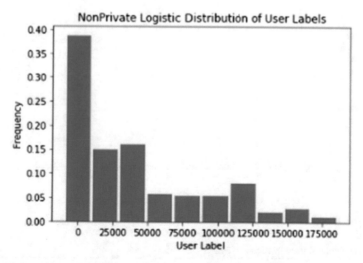

Fig. 13 Non-private data distribution

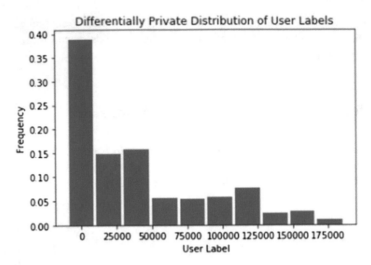

Fig. 14 Distribution after diff privacy epsilon = 0.8 applied

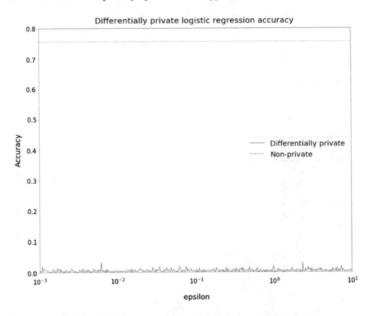

Fig. 15 Visualising the trade-off between accuracy and epsilon for this dataset

Table 3 Differential privacy logistic regression performance

Implementation	Baseline (epsilon = inf) (%)	Differential Privacy applied, epsilon = 0.8 (%)
Performance score	73.74	0.69
Privacy metric	26.26	99.31

6.2   Utility Metric

As epsilon approaches infinity it achieves 73.74% performance (with the least amount of added noise). This means that 26.26% of the data is non-discoverable. With differential privacy applied, 99.31% of the data is non-discoverable, and at this point it has minimal utility. This would then be counter-balanced by the removal of the "noise" at the receiving server.

6.3   Overhead Metric

As with k-means differential privacy, the overhead is dependent on the level of privacy vs the strain on bandwidth. For this review, the algorithm ran quite fast compared to the k-anonymity and dummy-based location application. There is far less data preparation required as well as the noise itself is a mathematical exponential variable rather than a geographically dependent variable.

7   Applying Dummy Based Location Anonymisation

A rudimentary algorithm was applied to test the anonymisation function of dummy-based location privacy. Dummy points were retrieved using "faker" Python module and applied to the dataset.

7.1   Privacy Metric

Using Logistic regression to predict the user location on the non-private and dummy location dataset, it would be difficult for a machine algorithm to decipher the location of the user. However, in Fig. 16, some of the purple data points might be in implausible locations when viewed from a human perspective. This could render some data points discoverable by human operators [8]. 72.87% privacy was obtained by using dummy locations, which is better than the baseline dataset privacy of 24.38% (Table 4).

7.2   Utility Metric

Provided the receiving server can remove the dummy points, the utility for spoofing/dummy locations would be high.

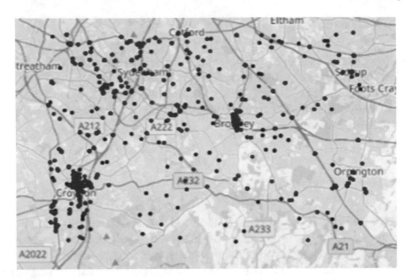

Fig. 16 Red points are the user locations, the purple points are the dummy locations

Table 4 Differential Privacy
Logistic Regression
Performance

Implementation	Non-private dataset (%)	Dataset with dummy locations (%)
Algorithm Performance	75.62	27.13
Privacy metric	24.38	72.87

7.3 Overhead Metric

Servers belonging to both the LBS and the user would be impacted by the added volumes of dummy users. However, in this rudimentary "experiment" the impact on the query runtime was minimal.

8 Results

We can now finally compare each of the different algorithms, in Table 5.

The baseline data with zero obfuscation, is followed by k-anonymity using l-diversity and t-closeness algorithms, then differential privacy with logistic regression. Finally, these are followed by the application of spoofing to the data. Each of these algorithms are compared on performance metrics privacy, utility, and overhead. The k-Means results are reported separately. The k-Means score is not comparable to the other algorithms in this review as the noise was applied to the latitude and

Table 5 Supervised learning algorithm performance results

	Algorithms	All algorithms	k-Anonymity			Differential Privacy	Dummy Locations (Spoofing)
Performance Metrics		(no privacy)	k-anonymity k=3	l-diversity k=3, l=5	t-closeness k=3, t=0.5	Logistic Regression epsilon = 0.8	faker
	Privacy	24~34%	73.92%	83.26%	83.33%	99.31%	72.87%
	Utility	Full utility.	Average utility. Would give reasonable recommendations.	Average utility. Would give reasonable recommendations.	Low to average utility. Depends on distance calculation used for closeness	Provided the Laplacian noise is removed for returning results, this would have full utility.	Provided the user server can remove the dummies, the data would have full utility.
	Overhead	None	Medium – High. Aggregating the data over large volumes could be costly			Low-High depending on level of privacy required	High

Table 6 Unsupervised Learning result—Differential Privacy by k-means clustering

Algorithm	Non-Private baseline	Differential Privacy with epsilon $= 0.1$
Privacy	Low	High
Utility	Full utility with defined stable clusters	Low utility with undefined changing clusters
Overhead	None	Low–High depending on level of privacy required

longitude. Although it's tempting to compare the privacy results directly, it is important to keep in mind that modifying any parameters such as k, l, t, or epsilon will yield different results. In utility, all algorithms except for k-anonymity t-closeness performed relatively well. The final comparison of overhead showed that any of the algorithms could incur an overhead cost based on a server bandwidth versus privacy trade-off. When taking all three-performance metrics into account, the most effective algorithm was the differential privacy using logistic regression (eps $= 0.8$) (Table 6).

9 Conclusion

Cyber security for location-based services can be accomplished using artificial intelligence techniques. Algorithms such as k-anonymity, differential privacy and location spoofing can effectively be applied to cloak user's sensitive data from malicious attacks. From our findings we can see that some algorithms do this more effectively than others.

This review exposed the challenge of balancing user privacy vs efficacy of the location-based service, highlighting the many variables and parameters. Whether it is through the choice of one or more AI algorithms or parameter management, there is room to improve and deliver the best outcome for all parties.

References

1. J. Hongbo, L. I. Jie, Z. Ping, Z. Fanzi, X. Zhu, A. Iyengar, Location Privacy-preserving Mechanisms in Location-based Services: A Comprehensive Survey. ACM Comput. Surv. **54**(1), 1–36 (2021). https://doi.org/10.1145/3423165
2. F. Fei, S. Li, H. Dai, C. Hu, W. Dou, Q. Ni, A K-Anonymity Based Schema for Location Privacy Preservation. IEEE Transactions on Sustainable Computing, Sustainable Computing, IEEE Transactions on, IEEE Trans. Sustain. Comput. **4**(2), 156–167 (2019). https://doi.org/10.1109/TSUSC.2017.2733018
3. D. Deutch, A. Frankenthal, A. Gilad, Y. Moskovitch, On Optimizing the Trade-off between Privacy and Utility in Data Provenance. arXiv:2103.00288 (2021)
4. Diffbot, Data provenance. Diffbot. https://blog.diffbot.com/knowledge-graph-glossary/data-provenance/

5. D. Slijepčević, M. Henzl, L. D. Klausner, T. Dam, P. Kieseberg, M. Zeppelzauer, k-Anonymity in Practice: How Generalisation and Suppression Affect Machine Learning Classifiers. arXiv: 2102.04763 (2021)

6. Q. Yu, Y. Luo, C. Chen, X. Ding, Outlier-eliminated k-means clustering algorithm based on differential privacy preservation. Appl. Intel. Int. J. Artif. Intell. Neural Netw Complex Probl. Solving Technol. Original Paper **45**(4), 1179 (2016). https://doi.org/10.1007/s10489-016-0813-z

7. K.-P. Lin, Privacy-preserving kernel k-means clustering outsourcing with random transformation. Knowl. Inf. Syst. **49**(3), 885–908, (2016). https://doi.org/10.1007/s10115-016-0923-2

8. T. Hara, A. Suzuki, M. Iwata, Y. Arase, X. Xie, "Dummy-Based User Location Anonymization Under Real-World Constraints," IEEE Access ; volume 4, page 673–687 ; ISSN 2169–3536, article in journal/newspaper 2016, doi: https://doi.org/10.1109/access.2016.2526060

9. Y. Du, G. Cai, X. Zhang, T. Liu, J. Jiang, An efficient dummy-based location privacy-preserving scheme for internet of things services. (in English), Information (Switzerland), vol. 10, no. 9 (2019). https://doi.org/10.3390/info10090278

10. Z. Wu, G. Li, S. Shen, X. Lian, E. Chen, G. Xu, Constructing dummy query sequences to protect location privacy and query privacy in location-based services. World Wide Web Internet Web Inf. Syst. Orig. Paper **24**(1), 25 (2021). https://doi.org/10.1007/s11280-020-00830-x

11. S. T. Mehedi, A. Anwar, Z. Rahman, K. Ahmed, Deep transfer learning based intrusion detection system for electric vehicular networks. Sensors (Basel, Switzerland), vol. 21, no. 14 (2021). https://doi.org/10.3390/s21144736

12. A. Anwar, A. N. Mahmood, M. Pickering, Data-driven stealthy injection attacks on smart grid with incomplete measurements (Lecture Notes in Computer Science). Springer Verlag (in English) (2016), 180–192

13. M. A. Husnoo, A. Anwar, Do not get fooled: Defense against the one-pixel attack to protect IoT-enabled Deep Learning systems. Ad Hoc Netw **122**, 102627 (2021). https://doi.org/10.1016/j.adhoc.2021.102627

14. L. Kristen, J. D. David, R. Raghu, Mondrian multidimensional k-anonymity. http://www.cse.iitb.ac.in/dbms/Data/Courses/CS632/Papers/kanon-lefevre-icde06.pdf, text (2006). https://ezproxy.deakin.edu.au/login?, https://search.ebscohost.com/login.aspx?direct=true&db=edsbas&AN=edsbas.5B63FCCD&site=eds-live&scope=site

15. IBM Differential Privacy Library: K-Means, IBM, diffprivlib (2021). https://diffprivlib.readthedocs.io/en/latest/modules/models.html#k-means

16. IBM Differential Privacy Library: Logistic Regression, IBM, diffprivlib (2021). https://diffprivlib.readthedocs.io/en/latest/modules/models.html#logistic-regression

17. S. Virupaksha, V. Dondeti, Anonymized noise addition in subspaces for privacy preserved data mining in high dimensional continuous data. Peer-to-Peer Netw. Appl. **14**(3), 1608 (2021). https://doi.org/10.1007/s12083-021-01080-y

18. J. Leskovec. Gowalla. SNAP. https://snap.stanford.edu/data/loc-gowalla.html. Accessed 14 May 2021

19. M. Fanourakis, Opportunistic multi-party shuffling for data reporting privacy, arXiv:2003.05273v1 [cs.CR] ed. University of Geneva, Switzerland: CUI, Quality of Life Lab 2020

20. L. Ninghui, L. Tiancheng, S. Venkatasubramanian, t-Closeness: Privacy Beyond k-Anonymity and l-Diversity, ed: IEEE, 106–115 (2007)

21. k-Anonymity, (2018). https://github.com/. https://github.com/Nuclearstar/K-Anonymity/blob/master/k-Anonymity.ipynb

22. L. Ninghui, L. Tiancheng, V. Suresh, Closeness: a new privacy measure for data publishing. IEEE Trans. Knowl. Data Eng. **22**(7), 943–956 (2010). https://search.ebscohost.com/login.aspx?direct=true&db=edb&AN=51119224&authtype=sso&custid=deakin&site=eds-live&scope=site

23. M. E. Andrés, N. E. Bordenabe, C. Palamidessi, K. Chatzikokolakis, Geo-indistinguishability: Differential privacy for location-based systems, pp. 901–914 (2013). https://doi.org/10.1145/2508859.2516735. https://search.ebscohost.com/login.aspx?direct=true&db=edselc&AN=edselc.2-52.0-84889037772&site=eds-live&scope=site

Chapter 10
Intelligent Radio Frequency Fingerprinting to Identify Malicious Tags in the Internet of Things

Biplob Ray, Grishma Khadka, Dimuth Lasantha, Varun Chandrappa, Jinho Choi, and Nemai Karmakar

Abstract The chapter has proposed a security framework for sensory tags on the Internet of Things (IoT) using intelligent Radio Frequency (RF) pattern analysis with machine learning to identify malicious tags. The RF Identification (RFID) system is an integral part of sensory tag based IoT applications. Due to limited computational capabilities on tags, traditional security primitives are not feasible for sensory tags, while there are existing lightweight security techniques that can be used to protect expensive RFID sensory tags, like active tags. However, less expensive tags, like passive tags, are incapable of using most of the existing lightweight security techniques. In addition, low-cost printed chipless sensory tags are incapable of using any existing security primitives at all due to no available computational and storage capabilities on board. Hence, implementing a holistic security provision to a hybrid sensory tag based IoT eco-system becomes complex or infeasible. The proposed security framework uses transmitted RF signals to counter security attacks, like malware injection, repudiation, and counterfeit, that are usually designed to infiltrated IoT system using malicious tag to address the security needs of all categories of sensory tags in the IoT. The proposed security framework extract features from RF data to prepare a fingerprint of the tag for its identity in association with other tags in the network, which is then used in machine learning for classification. It only needs to use computational and storage power of the reader to free the tags from computation and storage burden. The experiment on the proposed model shows that it can identify malicious tags with a high accuracy which is validated against several other machine learning techniques.

B. Ray (✉) · D. Lasantha · V. Chandrappa
School of Engineering and Technology, Centre for Intelligent Systems (CIS), Central Queensland University, Rockhampton, Australia
e-mail: b.ray@cqu.edu.au

G. Khadka · J. Choi
School of Information Technology, Deakin University, Burwood, VIC 3125, Australia

N. Karmakar
Department of Electrical and Computer System Engineering, Monash University, Clayton, VIC 3800, Australia

Keywords RFID · RF fingerprinting · Internet of Things · Chipless · Printable · Machine learning · Smart home · Malicious tags · Backscatter communication

1 Introduction

The Radio Frequency Identification (RFID) is a proven technology to uniquely identify smart objects on the Internet of Things (IoT) application [1]. The smart objects in IoT need to be identified using application generated unique keys to join the IoT network using communication protocols like Narrowband Internet of Things (NB-IoT) or Low-Power Wide Area Network (LoRaWAN) [2]. However, the application generated key exposes the IoT system to several vulnerabilities as an intruder can duplicate or change the application layer's unique key very easily to compromise the identity of the objects and/or the privacy of the owner [3]. The RFID based sensory tag is a high contender for an alternative to the application layer key for IoT objects. The recent development of RFID based sensory tag, lead to their reduced cost and size, makes it feasible to integrate RFID tags into smart IoT objects as their hardware-based unique identification [4]. There are three main categories of RFID sensory tags: active, semi-passive and passive. While active and semi-passive tags are expensive but possess higher computational and storage capability than their low-cost counterpart passive tags. Some IoT applications and their associated smart objects are mission critical, requiring expensive active tags; however, most everyday smart IoT objectives are not mission critical and require cost-effective passive tags to be viable. Therefore, the IoT eco-system consists of both active and passive sensory tags to identify smart objects. However, the recent developments of passive chipped and printed sensory, RFID tags constitute inexpensive, long-lasting, ubiquitous sensory tag units with unique identifiers [4] for smart objects connected to the internet. Smart materials combined with the passive RFID tags provide the snipping and remotely seeing capacity of tagged IoT objects. The passive chipped, and chipless sensory tag removes the high cost of integrating sensory technology with smart objects. Furthermore, the fully printable and passive tags that can even be printed on metals and bottles containing liquids, are resistant to extremely harsh environments and weather conditions [5]. The passive Radio Frequency Identification (RFID) based sensory tags can fill the gap in the IoT system domain by offering cost effective hardware identification and sensing solutions for smart objects to deploy in hostile environmental context/terrain [5]. The passive sensory tag technologies allow the integration of internet enabled smart tags [6], such as IoT enabled RFID based sensors, to objects in various application domains, like supply chain, and healthcare applications, to name a few.

Although sensory tag based unique key is more secure than application-based key for smart objects, there are still security concerns, like malware injection, repudiation, and counterfeit, that need to be addressed to make sensory tag based IoT system secure [1, 4, 7–9]. There are many lightweight cryptographic, authentication and security framework proposed in the literature which may be feasible to implement in the active or semi-passive sensory tags [1, 7–9] to secure sensory tag based IoT smart objects. However, due to being low computational and storage capable, the passive sensory tags, either non-printable or printable, are unable to take advantage of existing security techniques which are resource hungry [1, 2, 7–9]. Some existing work claimed to be feasible in non-printable sensory tag with Integrated Circuit (IC) [1]. For example, Ray et al. [8] have proposed security primitive, like malware detection, fully hosted and processed in the reader with minimal storage and process in the tag. But little or no security techniques exist that can protect the security of sensory tag based IoT systems, non-printable or printable, without putting the computational burden to the tag end. The on-board computational capability is non-existence in chipless RFID, printed or non-printed, based IoT systems. In addition, implementing of a security framework for a hybrid sensory tag based IoT system, mix of semi-passive and passive, is very complex or infeasible using existing techniques as tags must perform various levels of computation [4]. One of the key security concerns for passive RFID sensory tag is the infiltration of the IoT system using malicious tags mimicking genuine tags of the system [2, 4], which may be responsible for security attacks, like malware injection, repudiation, and counterfeit [4, 5]. For example, an adversary may attempt to inject simple and small malicious commands, like 'kill' or 'off', to compromise critical assets of the IoT application [8].

To device a standard security solution for all types of RFID sensory tag based IoT objects, this chapter has explored the signal patterns of RFID tags to the reader, using novel pattern analysis and intelligent machine learning to propose a security framework. The signal transmitted by RFID tags in the physical layer follow patterns based on their activities, position, environment, and duration of existence in the reader's interrogation zone. In this proposal, the patterns of the electromagnetic signal during interrogation by reader intelligently mined to extract features sets like distance and angle of electromagnetic signals relative to other tags. The proposed security framework aims to identify a malicious tag and their relative position using intelligent pattern analysis to stop security attacks that are initiated by introducing a malicious tag in the system. The proposed model requires computation and storage in the reader using a transmitted electromagnetic signal from a tag within the reading rage; therefore, the sensory tags require zero processing and storage. The contributions of this study are as follows.

- A standard security framework for sensory tag based IoT system using Radio Frequency (RF) signal of the sensory tags.
- Intelligent and novel pattern analysis of pre, during and post RF data transmissions for feature extraction and RF fingerprinting, which can reveal malicious tag and its relative location.
- A machine learning to classify tags within the reading range of the reader.
- Verification and validation of the proposed security framework to present its accuracy against other machine learning approaches.

2   Background

This section has detailed background and literature of RFID sensory tags and their security needs.

2.1   *Working Mechanism and Sensory Tag Evolution*

A reader spreads its electromagnetic signal to the power and interrogate the sensory tag within the reading rage, as illustrated in Fig. 1. Upon receiving the interrogation signal, the tag communicates to the reader and sends data based on the rules of the communication protocol, as illustrated in Fig. 1. The reader is connected to the backend to store and/or match the tags data.

Although conventional RFID tags have moved into mainstream applications, they still face technical challenges in large-scale adoption for the Internet of Things (IoT) applications. The first issue is the cost of the tag. The tag consists of an antenna and the application-specific integrated circuit (ASIC) to encode the information [10]. The design and assembling of the ASIC increase the fabrication cost in an integrated circuit (IC)-based tag. Due to the bulky structure of the tag, the printing of the tag on commercial paper or polymer substrate is impossible. Also, the IC-based tag is not able to withstand a cold temperature below zero and a high temperature above 40 °C [11].

The recent development of the sensory RFID opens new possibilities to tackle these issues [12, 13]. A sensory RFID consists of passive resonators on a single-layer structure. These passive microwave resonant elements without silicon chip act as identification bits. Each resonator forms a peak or null in the frequency response of the tag based on its property. The absence of a silicon chip significantly reduces the implementation cost in comparison to the chipped tag. Although the sensory RFID tags have definite cost advantages but are limited to the number of bits. For large

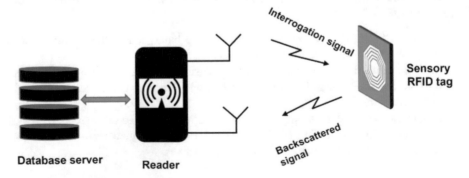

Fig. 1 Working mechanism of a passive sensory tag

data requirements, conventional RFID tags have upper levels but with a high price per tag. A sensory RFID tag can sustain in a high-temperature environment, whereas silicon chipped tag cannot. Besides being fully printable, it is a favorable solution to overcome the limitation of conventional tags for next-generation RFID technology. In a nutshell, the main advantages of a passive sensory RFID system in contrast to the conventional RFID system and optical barcode are explained below:

(a)   Robustness: It can be made fully printable with a commercially available printer and conductive ink, which can sustain its operation in a harsh environment.
(b)   Cost: It can significantly reduce the cost of production on a large scale as sensory tags do not possess onboard ASIC or any lumped components.
(c)   Size: The frequency domain backscattering mode based sensory RFID tag consists of resonating elements without the separate antenna. It meets the goal to make the tag of a credit card size with sufficient data for identification.
(d)   Spectrum Utilization: It avoids expensive licensing fee which works with low energy level and low power and large bandwidth technology suitable for smart retails. The low power makes it more feasible to use in radio frequency sensitive environments.

   Although the chipless RFID tags have many advantages, they cannot use conventional security primitives to protect themselves from security attacks due to non-existence computational and storage capabilities. In the next sub-section, the literature on sensory RFID tags is discussed in detail.

2.2   RFID Security Literature

The RFID based sensory tags are deployed on the user end, which are easily accessible by intruders for introducing malicious tags. Using a malicious tag an adversary may be able to generate attacks, like malware injection, repudiation, and counterfeit, for interrupting user end and backend IoT system components. Out of three different types, active, semi-passive and passive, of RFID sensory tags, the active tags have the best security protection available but can be vulnerable to malicious tags introduced by an inside attacker [14]. There are many existing security techniques and algorithms to ensure security requirements, like authenticity and confidentiality, for sensory tags, which require some computation in the tags. However, many passive tags, printable and non-printable, have minimal storage capability and are incapable of running the required computation in the tag. This literature review will cover existing techniques to secure passive sensory tags.

   The security of passive sensory tags can be mainly classified into two main categories: cryptographic and non-cryptographic [15, 16]. Within the cryptographic categories we have heavyweight [17–21], lightweight [22–29] and ultra-lightweight [30–35] cryptographic protocols. The heavyweight and lightweight categories are more applied for active and semi-passive tags, whereas ultra-lightweight primitives are for some passive RFID tags. The non-cryptographic security protocols are very few compared to the cryptographic technique, mainly based on machine learning or RF fingerprinting or a hybrid of both [16, 36–40]. Many non-cryptographic techniques, based on machine learning or RF fingerprinting, do not require computation and storage in the tag, making them very suitable for passive tags, both printable and non-printable, compared to cryptographic security techniques and frameworks. This section detailed ultra-lightweight [30, 31] cryptographic and non-cryptographic [16, 36–40] protocols or frameworks which are feasible in low-cost passive tags.

Ray et al. [30] have proposed a lightweight hybrid approach based on stenographic and watermarking to protect data confidentiality and integrity properties of tags data. The proposed lightweight hybrid approach [30] has used general characteristics of number theory, but the computation requires to be done in the tag. The technique does not protect many required security properties, like counterfeit and repudiation, and is not feasible for low cost chipless tags. Baashirah et al. [15], Aggarwal et al. [31], and Sharma et al. [32] have extensively analysed existing RFID security protocols that revealed several security weaknesses of existing ultra-lightweight security techniques to protect security properties, like malicious signal injection, repudiation, privacy, and counterfeit. In the light of the analysis of existing techniques, Rotter [33] have proposed a security protocol that requires XOR and bit-wise operation whereas Sharma et al. [32] proposal requires to perform XOR and hash function in the tag. Huang et al. [36] proposed a RFID mutual authentication based on the PadGen function in the ISO 18000-6C [34] protocol to protect the access password stored in the memory. The concept of their protocols is to secure the tag's access password before transmitting using PadGen with XOR or Modulo operation (MOD) in the tag. While the claim of these ultra-lightweight security protocols is always questionable as a motivated and resourceful adversary may be able to use known techniques to decode XOR or simple MOD operation to compromise claimed security. Mandel et al. [36] have proposed a lightweight and secure key storage scheme using novel Physical Unclonable Functions (PUFs) which requires special hardware in the tag. The authors [36] have claimed that their proposed PUF functions can restrict the change of adversaries even they are equipped with machine learning to analyse Radio Frequency (RF) transmissions.

In the non-cryptographic end, the proposals [16, 37–40] used a combinational of RF analysis and machine learning model to device a physical layer fingerprinting for the tag to protect it from security instructions. Bertoncini et al. [16] have proposed an RF fingerprinting technique that uses dynamic wavelet for feature extraction to be used by supervised pattern classification techniques to authenticate individual RFID tags with 99% accuracy without doing any computation in the tag. However, their implementation does not differentiate between genuine and malicious tags. Lakafosis et al. [37] have designed a reader to validate RF fingerprinting to identify the authenticity of tags called RF certificates of authenticity (RF-CoAs). This paper [37] has demonstrated the viability of RF fingerprinting for authenticity in real time using reader end communication only. Nouichi et al. [38] have proposed universal SDR (Software Defined Radio) based RF fingerprinting using statistical techniques only. The SDR RF fingerprinting [38] model is proposed to uniquely identify IoT devices using RF features which is validated in the paper [38] using mobile phone signals.

While Soltanieh et al. [38] have presented a detailed review of RF fingerprinting techniques from a signal extraction point of view, Zhang et al. [40] have proposed an RF fingerprinting technique to identify High Frequency (HF) RFID proximity cards using normalized power spectral density using the card's response. Soltanieh et al. [39] have identified that most of the RF fingerprinting techniques are based on the transient part of the signal. Therefore, to succeed, the transient signal must be extracted precisely. Soltanieh et al. [39] have also discussed handful of existing techniques using modulated part of the signal using classification algorithms. Zhang et al. [40] have concluded that as the feature extraction from RF can be implemented under a normal working state, it is practical and efficient to identify RFID tags using RF fingerprinting.

Based on the above review, it is evident that the cryptographic techniques are applicable for some RFID tags but not others, whereas non-cryptographic techniques, like RF fingerprinting, are applied to all types of RFID tags. Furthermore, the RF fingerprinting are practical and show a better result to protect security for low-cost RFID tags, like non-printed and printed passive tags, compared to cryptographic techniques.

3 Overview

This section detailed system model, dataset and algorithm used to develop the proposed security framework.

3.1 System and Attack Model

This chapter uses IoT-enabled smart homes that use sensory tags for object identification as illustrated in Fig. 2. The smart home has an IoT enabled master reader which can communicate using Long Range Wide Area Network (LoRaWAN) IoT protocol and Ultra-Wideband (UWB) transponder technology to communicate with sensory tags.

The master reader can also communicate with the internet via traditional communication like Wireless Fidelity (WiFi) or 5th generation mobile network (5G). The client reader, reports to the master reader, is used to boost LoRaWAN and UWB signal for the smart objects eco system of the smart home. Each smart objects in the smart home are enabled to communicate via Ultra-Wideband (UWB) or Low Power Personal Area Network (LPPAN).

As illustrated in Fig. 2, an adversary can introduce a malicious tag to the IoT system from outside the home by identifying a strategic location close to the master or client reader's signal. As explained previously, the malicious tag may be injecting false or garbage signals to either infiltrate the backend of the IoT system or create a Denial of Service (DoS), at least for a segment of the system. This has a huge impact

Fig. 2 Smart Home—system with attack model

on a family's day to day life in the smart home system where appliances or a secure document may be critical for their wellbeing or daily life. As illustrated in Fig. 2, an adversary may introduce a counterfeit document to confuse the IoT system to identify a secure document, therefore creating identity concern or panic to the stakeholders of the document. For example, in a remote verification scenario of a smart home, a counterfeit mortgage document may confuse the bank authority resulting in negative borrowing credit for the family member. In the final attack scenario, tempering with the physical signal of a sensory tag attached with an object may create repudiation issues while returning or handing over for costly maintenance as illustrated in Fig. 2.

Our proposed security framework can function in the master reader which will be able to do the RF fingerprinting to identify malicious tag and its relative position to eliminate the probability of these attacks.

3.2 Data Collection Process

In this section, as shown in Fig. 3, the experimental setup is used to show the backscatter signal pattern of the sensory RFID tag. A vector network analyser (VNA) is used as an RFID reader that reads the Scattering parameters (S-parameter) of the tag response. Initially, the reader transmits the signal using the 4×4 single patch antenna array. The reader recorded the backscatter signal of the tag (S11 parameter) within a specific frequency range placed at nearly 20–30 cm away. The collected dataset has phase and amplitude where the amplitude response variation is critical for a sensory tag's data transmission and identification. By varying amplitude signal an intruder may result insecurity when used for IoT applications. Thus, the chapter introduces the machine learning approach in analysing the RF tag pattern to overcome the challenge.

The measurement setup to obtain the dataset used the reader and tag as discussed in [41] where we obtained the ideal amplitude response of the frequency-based tag. The reading of the tag pattern is obtained in a real-time environment. The experiments assume the different sources of interferences such as antenna coupling, structural mode of the tag, interference from neighbor radios, echo from static and moving objects and supports which hold the sensory tag. The multiple measurements of the three different tags are performed in a robust environment to compute the backscatter pattern analysis using machine learning. To feed the machine learning model, we use the Gaussian copula-based framework to generate synthetic data. Here, the model is developed, which contains all the possible dependencies in a multivariate Gaussian

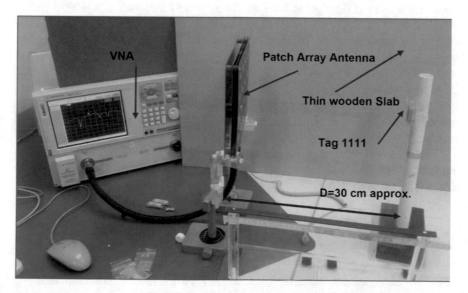

Fig. 3 Experimental setup for measuring backscatter tag signal pattern using the E8361A VNA and a 4×4 array antenna

Table 1 Sample phase (∅) data points

	4	4.05	4.1	4.15	4.2	4.25
1	−163.5435	176.278	156.0743	137.7088	117.4725	94.7983
2	−166.788	173.8937	153.248	133.5383	115.7608	92.5318
3	−166.9087	173.9921	153.1324	134.496	115.2576	92.2135

Table 2 Sample amplitude (A) data points

	4	4.05	4.1	4.15	4.2	4.25
1	−36.8449	−36.0962	−36.0657	−35.8132	−34.4963	−34.7369
2	−26.4173	−25.844	−25.378	−24.9555	−24.5635	−24.0979
3	−26.401	−25.6602	−25.2122	−24.7549	−24.215	−23.7247

distribution. The data generated using the model is statistically alike to the source data with dependence across variables. For our analysis, the synthetic data is generated for three different tags to obtain the desired subsets within the given frequency ranges, which is expected to improve the performance and provide more learning flexibility.

3.3 Dataset Explained

Each tag was read through at least 88 times in 254 sub-channels with different frequencies initiating from 3.3 GHZ and incriminating by 0.025 up to 9.625 GHz during the experiment. In Tables 1 and 2, we have presented some sample data points of three iterations. Table 1 contains tags phase (∅) data for three reading of 4–4.25 GHz frequency groups. For example, the first reading of 4.15 GHz indicates 137.7088^0 phase.

The Table 2 contains tags amplitude (A) data for three readings of 4–4.25 GHz frequency groups. For example, the first reading of 4.15 GHz indicates −35.8132 dB amplitude.

These phase (∅) and amplitude (A) are spread across 254 sub-channels for 88 iteration which are used to create some more synthetic data of phase (∅) and amplitude (A) as explained Sect. 3.2.

3.4 Novel Techniques Used for Intelligent RF Fingerprinting

We have calculated, centroids, distance, and angle using the RF data, amplitude (A) and phase (∅) for 254 sub-channels within 3.3–9.625 GHz, collected during tags transmission to the reader, as detailed below. These three features have helped us to prepare features for intelligent RF fingerprinting.

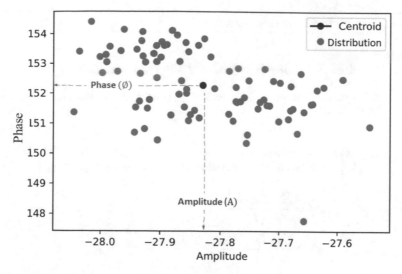

Fig. 4 Centroid, ($f_{sub}(\mathbb{C}_T)$)of sub-channel frequency (f_{sub}) of tag T

Centroid: It is the central point (or mean point) between amplitude (A) and phase
($\varnothing$) in a sub-channel frequency. To find the central point, we have used Eqs. (1) and
(2) to calculate average amplitude, $A_T^{\mathbb{C}}$, and phase, $\varnothing_T^{\mathbb{C}}$, of a sub-channel frequency
as detailed in Algorithm 1 which presents the process of calculating centroid ($\mathbb{C}_T$)
of a tag for all sub-channel frequencies one by one. The centroid of a sub-channel
frequency ($f_{sub}(\mathbb{C}_T)$)is illustrated in Fig. 4 by distributing amplitudes and phases
on a plane, represented using a green circle, where the x-axis has plotted amplitude
and the y-axis has plotted phase. The green circle represents, the meeting point of
amplitudes, $A_T^{\mathbb{C}}$, and phase, $\varnothing_T^{\mathbb{C}}$ on a distributed plane, the centroid, ($f_{sub}(\mathbb{C}_T)$)of
sub-channel frequency (f_{sub}) as illustrated in Fig. 4.

Algorithm 1 Finding the centroid for a tag

Input:

Iterations (i) ReadNo (R): $(1 < i <= R)$ i increments by reading number of the tag

Sub-channel frequency (f_{sub}): $(3.3 < f_{sub} <= 9.625)$ and f_{sub} increments by 0.025 GHZ

$f_{sub}(A_T^i)$ = Amplitude of i_{th} iteration of sub-frequency f_{sub} of tag T

$f_{sub}(\emptyset_T^i)$ = Phase of of i_{th} iteration of sub-frequency f_{sub} of tag T

$$A_T^C = \frac{\sum_{i=1}^{R} f_{sub}(A_T^i)}{i} \qquad (1)$$

$$\emptyset_T^C = \frac{\sum_{i=1}^{R} f_{sub}(\emptyset_T^i)}{i} \qquad (2)$$

Output: Centroid of a Sub-channel frequency $| f_{sub}(C_T) = (A_T^C, \emptyset_T^C) |$

Distance: In three-dimensional coordinate systems made of sub-channel (f_{sub}) and two centroids' values of two tags, the distance, $\mathcal{D}$, is calculated using Pythagorean theorem in Eq. (3) below where x, y, and z represent coordinate of three values of three-dimensional coordinates. For example, to get the distance between T_1 and T_2, the coordinate values of centroids, (C_{T_1}), (C_{T_2}) are considered in the plain of sub-channel frequency, f_{sub}, as x, y and z respectively.

$$\mathcal{D}(T_1, T_2) = \sqrt{(x_2 - x_1)^2 + (y_2 - y_1)^2 + (z_2 - z_1)^2} \qquad (3)$$

In Fig. 5, we have presented distance of three tags at 4GHz sub-channel where x-axis is presenting amplitude and y-axis is presenting phase centroid values.

Angle: The angle between two tags can be calculated using Eq. (4) in a three-dimensional coordinate system of a sub-channel (f_{sub}) as illustrated in Fig. 5.

$$\Lambda(T_1, T_2) = \arccos\left(\frac{x_1 x_2 + y_1 y_2 + z_1 z_2}{\sqrt{(x_1^2 + y_1^2 + z_1^2).(x_2^2 + y_2^2 + z_2^2)}}\right) \qquad (4)$$

In Eq. 4, the angle between T_1 and T_2 is represented by Λ and pair of coordinate values of (C_{T_1}), (C_{T_2}), and f_{sub} is considered as x, y and z respectively.

As illustrated in Fig. 5, distance and angle are corelated to multiple tags which can help to identify the relative position of a malicious tag.

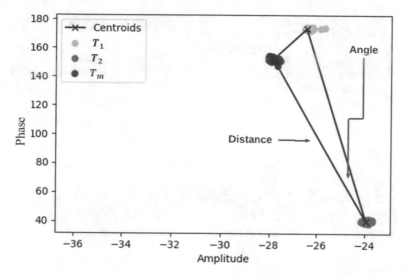

Fig. 5 Distance and angle between (T_1, T_2 and T_m) tags at 4 GHz sub-channel

3.5   *Gradient Boosted Machine (LGBM)*

Due to the nature of our RF data and further use of the RF fingerprinting on continuous live data, the proposed security framework used a boosting algorithm. There are three different types of boosting algorithms that are in trend in the industry. (i) Extreme Gradient Boosting (XGBoost) (ii) Light Gradient Boosted Machine (LGBM) and (iii) CatBoost. Our proposed framework has used LGBM to support continuous data, reduce the memory footprint, support parallel execution in learning to increase the model's speed, and perform very well on a large data set [42]. The decision tree based Light GBM is used in solving many regressions and classification problems [43]. The Light GBM is fast, and it uses less memory in computation. As shown in Fig. 6, LightGBM uses a histogram in creating a decision tree. For the given training data set, the Light GBM creates several bins to group the data into a different number of bins. Using this bin data histogram is created. The Light GBM uses this histogram to create the decision tree [44].

The Light GBM split the nodes leaf wise whereas the other model split the nodes based on the level. Figure 7 shows the leaf-wise splits in a decision tree.

The XGboost splits the nodes based on the level wise hence it grows horizontally whereas Light GBM splits the node leaf wise and grows vertically [10]. The leaf split reduces the loss compared to the level wise split hence Light GBM, achieves more accuracy in prediction compared to other boosting algorithms. The vertical splits might result in an overfitting problem, but it can be controlled by specifying the max depth option provided by Light GBM.

The LightGBM uses Eq. 5 to train the model where in y_i represents dependent variable, $f_c(x_i)$ represent current decision tree residual error, $f_{(c-1)}(x_{i-1})$ is the

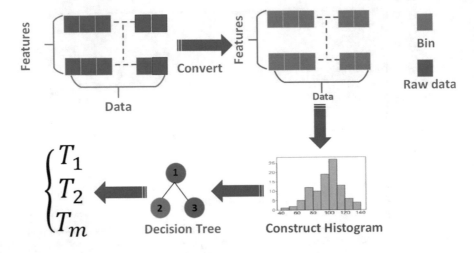

Fig. 6 Histogram based decision tress [44]

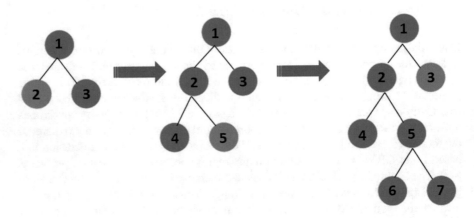

Fig. 7 LightGBM leaf wise tree growth

previous decision tree residual error which goes as input to the subsequent decision tree and L is loss function. The decision tree formation will continue until the residual error is zero or gain is high or user specified the maximum number of trees.

$$\Gamma_c = \sum_{i=1}^{N} L(y_i, f_{c-1}(x_i) + f_C(x_i))$$ (5)

4 Proposed Security Framework and Experiment Result

The proposed framework in this chapter has three main segments as illustrated in Fig. 8. In segment one, we collect data from Vector Network Analyzer (VNA) via reader module as detailed in Sects. 3.2 and 3.3. In segment two, the intelligent pattern analysis is performed to do RF fingerprinting features of the tags using techniques detailed in Sect. 3.4.

As illustrated in Fig. 8, the RF data is analysed in categories like pre, duration and post data transmission for identifying the centroid, $C(A, \varnothing)$, between amplitude (A) and phase ($\varnothing$) for each tag. The centroid is then used to calculate the distance ($\mathcal{D}$) and the angel (Λ) to create an RF fingerprint, RFF_T, of a tag as detailed in Sect. 5.1. The features pair distance ($\mathcal{D}$) and the angel (Λ) along with sub-channel frequencies, f_{sub}, passed to last segment of the framework to classify tags dynamically using machine learning as illustrated in Fig. 8. The framework usages computational capability of the master reader of our system model illustrated in Fig. 2 and detailed in Sect. 3.1 to identify malicious tag. The system first creates the RF signature of the genuine tags when initially deployed and stored in the backend. Thereafter, each new tag gets RF fingerprinted in real-time to identify the entry of a malicious tag within the reader's range proactively before the data transmission phase. The RF fingerprinting process extracts unique features between neighbor tags to determine their relative associativity, which helps identify possible relations between two tags and the relative location of a malicious tag ID within the reader's range. The machine learning classification of tag type based on unique features is a dynamic process to make the identification scalable and feasible for real time applications like the system model in Fig. 2.

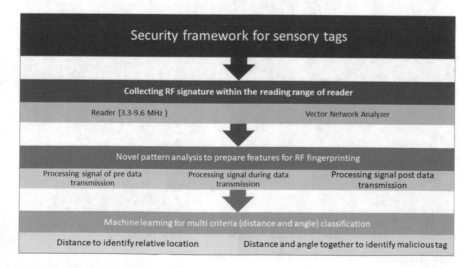

Fig. 8 Proposed security framework

5   Experiment Result

In this section, we detail the experiment result which is followed by discussion in
Sect. 6.

5.1   Intelligent RF Fingerprinting

The process of RF fingerprinting analysis is illustrated in Fig. 9 that shows before
starting the process of RF fingerprinting, we have analysed the amplitude (A) of
collected frequency as illustrated in Fig. 10.

The tag transmitting the data within the frequency range from 4.5 to 5.75 GHz
sub-channel frequencies is shown in Fig. 10. Therefore, we have considered that
all sub-channel frequencies before 4.5 GHz are in the pre-data transmission group,
whereas all sub-channel frequencies after 5.75 GHz are in post data transmission
group.

These leaves 4.5–5.75GHz sub-channel frequencies are in during the data transmission group. It is critical to identify the malicious tag just before start transmitting
the data as well as during and after the transmission if not identified earlier time.

As illustrated in Fig. 9, we initially calculate centroid,$C(A, \varnothing)$, using detail in sub-
Sect. 3.4. In Figs. 11 and 12, the centroid values pair and sub-channel frequencies of
three tags,$T_1 T_2 T_m$, are plotted in a 3D plain during (range from 4.575 to 4.775 GHz)

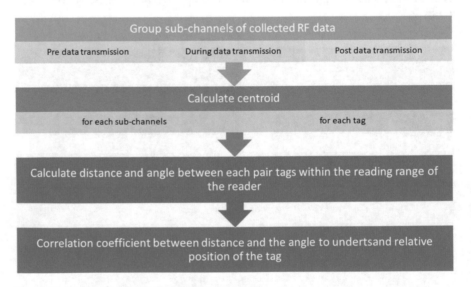

Fig. 9 The process flow of intelligent RF fingerprinting

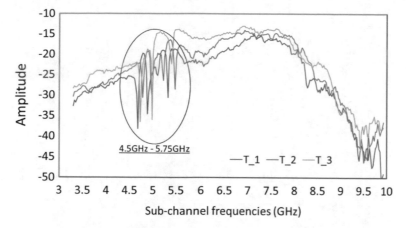

Fig. 10 Data transmission sub-frequencies

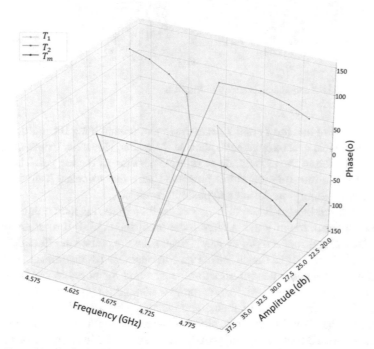

Fig. 11 Centroids of sub-channels during transmission

and post (range from 5.575 to 5.775 GHz) data transmission respectively where X-axis presents amplitude(A), Y-axis plotted phase ($\varnothing$) and Z axis plotted sub-channel frequencies(f_{sub}).

The illustration in Figs. 11 and 12 does not give us definite RF fingerprinting of tags, but it clearly shows that centroid of T_1 is close to T_m compared to T_2. This is

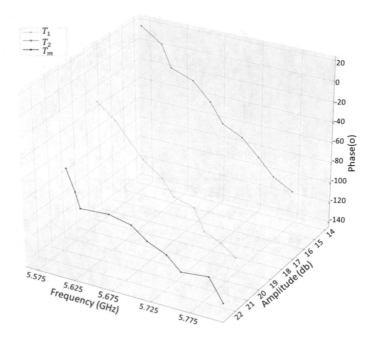

Fig. 12 Centroids of sub-channels after transmission

a clear indication that the T_1 and T_m may have similar/same tag ID and/or relatively close location in the electromagnetic plain of the reader. However, it does not provide a clear indication of the association or relative position of the tag yet. The centroid plot of pre data transmission is not presented here due to space limitation, but it shows similar findings as during and posts transmission.

Therefore, it is hard to distinguish the pattern for identifying their unique signature and association to identify the relative location of the tag. To further analysis, the centroids of all tags for each sub-channel frequencies are plotted in a Euclidean space, as detailed in Sect. 3.4 and illustrated in Fig. 5 to calculate distance and angle between tags. The process continues for all sub-channels and for possible combinations of the tags. In our experiment, we have calculated distance ($\mathcal{D}$) and angel (Λ) between $T_1 - T_m$, $T_1 - T_2$ and $T_2 - T_m$ to understand their signature, associativity, and relative position. In Figs. 13 and 14, we have plotted distance ($\mathcal{D}$) with associated sub-channel frequencies in a 2D plot for pre and post data transmission where the x-axis presented the sub-channel frequencies, and the y-axis plotted the distance.

As we can see from Figs. 13 and 14, the distance between $T_1 - T_m$ has consistent low value before and after the transmission. This means malicious tag T_m is closer to T_1 compared to any other tags on the plain. The distance plot during transmission has presented similar observations but due to space limitation not presented here.

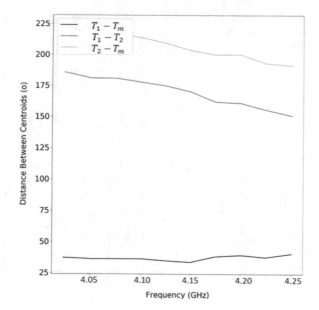

Fig. 13 Distance between Centroids before the transmission

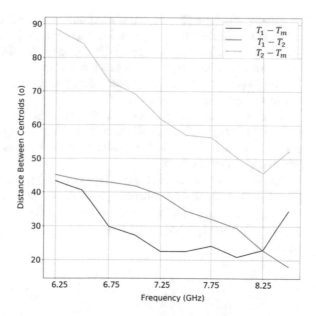

Fig. 14 Distance between Centroids after the transmission

Like distance, the angle (Λ) and sub-channel frequencies are plotted in a 2D plot for pre, during and post data transmission. Figure 15 has illustrated angle over sub-channel frequencies of post data transmission group where the x-axis presented the sub-channel frequencies, and the y-axis plotted the angle.

Fig. 15 Angle between
Centroids after the
transmission

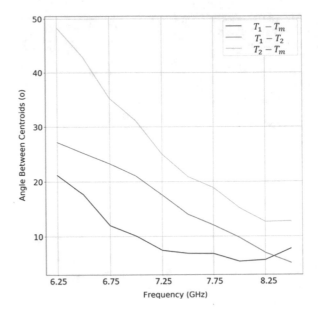

The angle between $T_1 - T_m$ has consistent low value compared to other pairs which shows that the malicious tag T_m is inclined towards T_1. Based on our experiment setup, we know the T_m and T_1 are stored with the same tag ID, therefore, counterfeit tags will have the lowest angle and distance between them on a reader's reading plain. The angle plot pre and during transmission have presented similar observations but due to space limitation not presented here.

Based on experiment data, the lower distance and angle values between two tags tells us they are closely located and/or hold the same tag ID. Therefore, it is evident that the distance and angle features can provide the associativity and the relative position between two tags.

Finally, the RF fingerprinting of a tag is derived using Eq. (6) where we have used all values throughout the frequency spectrum. As derived in Eq. (6), the RFF_T value is always calculated relative to a neighbor tag $\widehat{T}$ where centroid belong to the tag T but distance and angle is calculated relative to two tags as expressed by $T - \widehat{T}$ in the equation.

$$\begin{Bmatrix} T \\ \widehat{T} \end{Bmatrix} RFF_T = \sum_{f_{sub}=3.3}^{f_{sub}=9.675} (\complement, \Lambda, \mathcal{D})_{f_{sub}} \tag{6}$$

Finally, to validate the confidence level of our findings from RFF_T, we have calculated Correlation Index (CI) between the distance and angle of tag pairs using an Equation adapted from [45]. For example, let's consider $T_1 - T_m$, first, calculate distances and angles for all sub-channel frequencies for the pair, then calculate the correlation index among them.

Table 3 Correlation coefficient values

Tag pair	CI
$T_1 - T_m$	93.83%
$T_2 - T_1$	**82.65% (baseline value)**
$T_2 - T_m$	72.73%

Table 3 presents the CI value of three tag pairs where the highest correlation coefficient is for $T_1 - T_m$ pair which gives us the confidence to accept our previous finding that T_m is a malicious entry which is closely linked with T_1 by tag or location or both. The CI values between $T_2 - T_m$ and $T_2 - T_1$ are always low whereas angle and distance are always high compared to $T_1 - T_m$ pair. It is important to note that, the CI value between $T_2 - T_m$ is very low whereas between $T_2 - T_1$ is moderate. This gives us the confidence to accept CI value between $T_2 - T_1$ as a baseline value for this experiment.

To make this novel RF fingerprinting more adaptive and scalable, we have used machine learning to identify T_m using angle and distance features. The results of machine learning experiment are detailed below.

5.2   Machine Learning to Identify the Malicious Tag and Relative Position

From Sect. 5.1, distance ($\mathcal{D}$) and angel (Λ) values along with their sub-channel frequencies (f_{sub}) of each tag are used as input x_i in Eq. (5) of the LightGBM algorithm detailed in Sect. 3.5.

$$x_i = \begin{bmatrix} D_{f_{sub}1} & \cdots & D_{f_{sub}i} \\ \Lambda_{f_{sub}1} & \cdots & \Lambda_{f_{sub}1} \\ f_{sub}1 & \cdots & f_{sub}i \end{bmatrix} \tag{7}$$

The Eq. (7) shows the input (x_i) matrix whereas the Eq. (8) presents the output classes of the dataset passed to the LightGBM model to classify tags. We have randomised our entire dataset, which is then split into 70% training and 30% testing set to train and test the LightGBM algorithm. The experiment has used various boosting types like Gradient boosting Decision Tree (GBDC), Gradient-based One-Side Sampling (GOSS), and Dropouts meet Multiple Additive Regression Trees (DART).

$$y_i = \begin{cases} T_1 \\ T_2 \\ T_m \end{cases} \tag{8}$$

Fig. 16 Accuracy variation
of the proposed model
while using
different boosting techniques

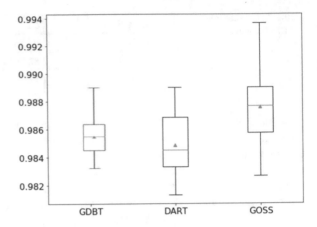

Based on the test result, the boosting type of GROSS performed very well with an accuracy of 98.8%, whereas GDBT and DART performed with an accuracy of 98.7 and 98.5, respectively as illustrated in Fig. 16.

To validate the accuracy of the LightGBM model, ten stratified K-folds cross validation is created which has processed the cross validation by swapping training and testing set as illustrated in Fig. 17. The cross-validation result shows that the LightGBM model has performed well with a mean accuracy of 98.76% with the least accuracy of 96.4%.

Furthermore, the LightGBM performance is evaluated by changing the hyperparameters like the number of decision trees, the tree's depth, and different boosting types. For decision tree, we have varied tree size, 10, 50, 100, and 500, which have achieved an accuracy of 96.5%, 98.4%, 98.7%, and 98.8%, respectively, as presented in Table 4. This result shows that the decision tree number has a linear relationship with the accuracy of the LightGBM algorithm.

The proposed LightGBM model also evaluated by varying the depth of the decision tree from 0 to 10 as illustrated in Fig. 18 where the model achieved best performance is 98.6%, and the least performance is 93.2%. The result shows that an increase in the depth of the tree, until 4, has increased the accuracy of the LightGBM model sharply and got steady afterwards as illustrated in Fig. 18.

Based on the results above the LightGBM model is identifying malicious and clean tag with very high accuracy based on features identified by intelligent pattern analysis in Sect. 5.1.

6   Discussion

It is evident from the experiment result; the proposed framework is very efficient to extract the RF fingerprint of the tag. By comparing the fingerprinting feature sets, centroid, angle, distance, and correlation coefficient, it can clearly identify the

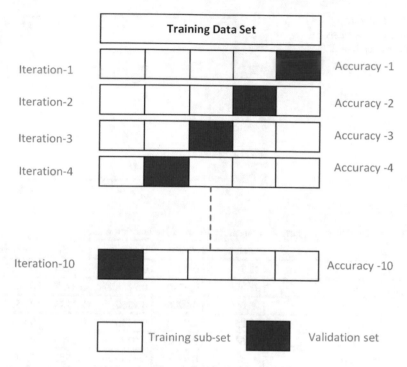

Fig. 17 Flow chart of Ten-fold cross validation

Table 4 Increase in number of Decision Tree cross validation accuracy

Experiment type	Average accuracy in 10- fold	Standard deviation
Increasing number of Decision tree	96.4	0.003
	98.4	0.002
	98.6	0.002
	98.7	0.002

malicious tag position. As each tag can be presented with an RF fingerprint in the reading range of the reader, the IoT system can ensure non-repudiation by verifying the signature of the tag attached to the smart object. As we can see in the experiment, the tag with a counterfeit tag ID will have a close position with a tag with same tag ID, therefore distance ($\mathcal{D}$) and angel (Λ) will be smaller, whereas the correlation coefficient value will be higher than the baseline value identified in Table 3. A Large deviation of CI from the baseline value means a malicious tag as the RF signal of the malicious tag will be very different from the genuine tag pool of the network. The proposed machine learning is ensuring the identity of a tag based on RF fingerprinting features of the tags. By comparing relative distance ($\mathcal{D}$) and angel (Λ) of a specific tag, the system can derive the relative location of the tag within the reading range of the reader.

Fig. 18 LightGBM
performance based on the
increasing depth of the tree

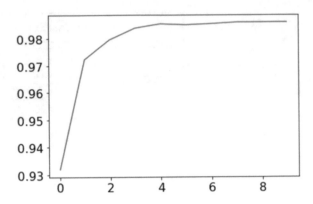

Table 5 performance comparison of our proposed machine learning model

Algorithm	Accuracy	Recall	Precision	RMSE	NRMSE
LGDBM	98.7	0.9075	0.9876	0.1114	0.0553
BaggingClassifier	97.08	0.9710	0.9711	0.1759	0.087
Xgboost	98.6	0.9906	0.9860	0.1183	0.0589
Linear Discriminant analysis	94.75	0.9474	0.9480	0.2292	0.1140
Quadratic discriminant analysis	92.68	0.9268	0.9278	0.2706	0.136
Extra trees classifier	97.59	0.9759	0.9759	0.1553	0.0773

We have also compared the performance of our proposed statistical machine learning model with some existing Neural networks and Deep learning-based models on the features like accuracy, recall, precision, RMSE, and NRMSE, as listed in Table 5. The proposed LGDBM framework has the highest accuracy and lower error rate, as evident based on data of Table 5.

7   Conclusion

This chapter has proposed a security framework using intelligent RF analysis and machine learning to identify malicious tags and their relative location to counter attacks such as counterfeit, repudiation, and malicious signal injection in sensory tag based IoT system. The proposed framework has used novel pattern analysis to calculate RF fingerprinting of tags in the relative existence of other tags in the IoT network. The RF fingerprinting values let the security framework calculate the correlation coefficient of the tag with other tag to identify type of instruction by comparing it with the baseline. The framework's machine learning model ensures adaptive identification of the tag and tag types with 98.7% accuracy based on extracted RF fingerprinting features. The proposed security framework can secure all types of tags, including printed chipless RFID base sensory tags in IoT systems. In our future work, the proposed framework will be tested with a large dataset in a live network of 50 or more sensory tags to improve and validate the baseline coefficient and knowledge of the machine learning model. We also aim to develop a mathematical model and dashboard to determine and visualise specific location of the malicious tag using distance ($\mathcal{D}$) and angel (Λ) feature set.

References

1. B.R. Ray, M. Chowdhury, J. Abawajy, Critical analysis and comparative study of security for networked RFID systems, in *2013 14th ACIS International Conference on Software Engineering, Artificial Intelligence, Networking and Parallel/Distributed Computing* (2013), pp. 197–202. https://doi.org/10.1109/SNPD.2013.41
2. X. Chen, M. Lech, L. Wang, A complete key management scheme for LoRaWAN v1.1. Sensors (Basel) **21**(9), 2962 (2021). https://doi.org/10.3390/s21092962
3. M. Frustaci, P. Pace, G. Aloi, G. Fortino, Evaluating critical security issues of the IoT world: present and future challenges. IEEE Internet Things J. **5**(4), 2483–2495 (2018). https://doi.org/10.1109/JIOT.2017.2767291
4. J. Grosinger, W. Pachler, W. Bosch, Tag size matters: miniaturized RFID tags to connect smart objects to the internet. IEEE Microwave Mag. **19**(6), 101–111 (2018). https://doi.org/10.1109/MMM.2018.2844029
5. R.E. Azim, R. Koswatta, N.C. Karmakar, P. Kalansuriya, Chipless radio frequency identification reader signal processing (John Wiley & Sons, Incorporated, Hoboken, 2016)
6. A. DiCarlofelice, E. DiGiampaolo, P. Tognolatti, mm-wave chipless RFID tag for healthcare applications, in *XXXIIIrd General Assembly and Scientific Symposium of the International Union of Radio Science* , pp. 1–3 (2020). https://doi.org/10.23919/URSIGASS49373.2020.9232382
7. B. Ray, S. Huda, M.U. Chowdhury, Smart RFID reader protocol for malware detection (IEEE, 2011), pp. 64–66
8. B. Ray, M.U. Chowdhury, T. Pham, Mutual authentication with malware protection for RFID system, in *Annual International Conference on Information Technology Security (ITS 2010)* (Singapore, 2010), pp. I-24–I-29
9. B.R. Ray, J. Abawajy, M. Chowdhury, Scalable RFID security framework and protocol supporting Internet of Things, in *Computer networks*, vol. 67 (Amsterdam, Netherlands: 1999), pp. 89–103. https://doi.org/10.1016/j.comnet.2014.03.023
10. R. Want, An introduction to RFID technology. IEEE Pervasive Comput. **5**(1), 25–33 (2006). https://doi.org/10.1109/MPRV.2006.2
11. Rubayet-E-Azim, Collision, data recovery and localisation in Chipless RFID. Monash University. Thesis (2017). https://doi.org/10.4225/03/58abb7c23596b
12. A. Attaran, R. Rashidzadeh, Chipless radio frequency Identification tag for IoT applications. IEEE Internet Things J. **3**(6), 1310–1318 (2016). https://doi.org/10.1109/JIOT.2016.2589928
13. T. Athauda, N.C. Karmakar, The realization of chipless RFID resonator for multiple physical parameter sensing. IEEE Internet Things J. **6**(3), 5387–5396 (2019). https://doi.org/10.1109/JIOT.2019.2901470
14. M.U. Chowdhury, R. Doss, B. Ray, S. Rajasegarar, S. Chowdhury, IoT insider attack-Survey (Springer International Publishing, Cham , 2020), pp. 28–41
15. R. Baashirah, A. Abuzneid, Survey on prominent RFID authentication protocols for passive tags. Sensors (Basel) **18**(10), 3584 (2018). https://doi.org/10.3390/s18103584
16. C. Bertoncini, K. Rudd, B. Nousain, M. Hinders, Wavelet fingerprinting of radio-frequency identification (RFID) tags. IEEE Trans. Ind. Electron. (1982) **59**(12), 4843–4850 (2012). https://doi.org/10.1109/TIE.2011.2179276

17. J. Wang, C. Floerkemeier, S.E. Sarma, Session-based security enhancement of RFID systems for emerging open-loop applications. Pers. Ubiquit. Comput. **18**(8), 1881–1891 (2014). https://doi.org/10.1007/s00779-014-0788-x
18. ISO. ISO/IEC DIS 9798–2. In Information Technology-Security Techniques-Entity Authentication—Part 2: Mechanisms Using Authenticated Encryption (International Organization for Standardization, Geneva, Switzerland, 2017)
19. E.-K. Ryu, D.-S. Kim, and K. Yoo, "On Elliptic Curve Based Untraceable RFID Authentication Protocols," Proceedings of the 3rd ACM Workshop on Information Hiding and Multimedia Security, 2015.
20. R. Songhela, M. Das, Yet another strong privacy-preserving RFID mutual authentication protocol, in SPACE (2014)
21. J.-S. Chou, An efficient mutual authentication RFID scheme based on elliptic curve cryptography. J. Supercomput. **70**(1), 75–94 (2014). https://doi.org/10.1007/s11227-013-1073-x
22. B. Ray, M.U. Chowdhury, Enhanced RFID mutual authentication scheme based on shared secret information, in *CAINE 2010: Proceedings of the ISCA 23rd International Conference on Computer Applications in Industry and Engineering* (International Society for Computers and their Applications (ISCA), 2010), pp. 208–213
23. B. Ray, S. Huda, M.U. Chowdhury, Smart RFID reader protocol for malware detection, in *2011 12th ACIS International Conference on Software Engineering, Artificial Intelligence, Networking and Parallel/Distributed Computing, 2011*, pp. 64–69. https://doi.org/10.1109/SNPD.2011.41
24. X. Chen, T. Cao, J. Zhai, Untraceability analysis of two RFID authentication protocols. Chin. J. Electron. **25**, 912–920 (2016)
25. H.-Y. Chien, C.-W. Huang, A lightweight authentication protocol for low-cost RFID. Journal of signal processing systems **59**(1), 95–102 (2010). https://doi.org/10.1007/s11265-008-0281-8
26. ISO. ISO/IEC 15693–2:2006, In Identification Cards-Contactless Integrated Circuit Cards-Vicinity Cards—Part 2: Air Interface and Initialization (International Organization for Standardization, Geneva, Switzerland, 2006)
27. M. Burmester, J. Munilla, Lightweight RFID authentication with forward and backward security. ACM Trans. Inf. Syst. Secur. **14**(1), 1–26 (2011). https://doi.org/10.1145/1952982.195 2993
28. L. Zilong, L. Dongsheng, L. Lun, L. Hui, Y. Zhenqiang, Implementation of a new RFID authentication protocol for EPC Gen2 standard. IEEE Sens. J. **15**(2), 1003–1011 (2015). https://doi.org/10.1109/JSEN.2014.2359796
29. N. Haifeng, E. Taqieddin, S. Jagannathan, EPC Gen2v2 RFID standard authentication and ownership management protocol. IEEE Trans. Mob. Comput. **15**(1), 137–149 (2016). https://doi.org/10.1109/TMC.2015.2412933
30. B.R. Ray, M. Chowdhury, J. Abawajy, Hybrid approach to ensure data confidentiality and tampered data recovery for RFID tag. Int. J. Netw. Distrib. Comput. (Online) **1**(2), 79 (2013). https://doi.org/10.2991/ijndc.2013.1.2.2
31. R. Aggarwal, M.L. Das, RFID security in the context of "internet of things", in *Proceedings of the First International Conference on Security of Internet of Things, Kollam, India, 17–19 August 2012* (2012)
32. V. Sharma, A. Vithalkar, M. Hashmi, Lightweight security protocol for chipless RFID in Internet of Things (IoT) applications, in *2018 10th International Conference on Communication Systems & Networks (COMSNETS), 2018*, pp. 468–471. https://doi.org/10.1109/COMSNETS.2018.8328246
33. P. Rotter, A framework for assessing RFID system security and privacy risks. IEEE Pervasive Comput. **7**(2), 70–77 (2008). https://doi.org/10.1109/MPRV.2008.22
34. ISO. ISO/IEC 18000–6:2013, In Information Technology-Radio Frequency Identification for Item Management—Part 6: Parameters for Air Interface Communications at 860 MHz to 960 MHz General (International Organization for Standardization, Geneva, Switzerland, 2013)
35. Y.-J. Huang, W.-C. Lin, H.-L. Li, Efficient implementation of RFID mutual authentication protocol. IEEE Trans. Ind. Electron. (1982) **59**(12), 4784–4791 (2012). https://doi.org/10.1109/TIE.2011.2178215

36. M.-D. Yu, D. M'Raihi, R. Sowell, S. Devadas, Lightweight and secure PUF key storage using limits of machine learning, in *Cryptographic Hardware and Embedded Systems–CHES 2011*, (Springer, Berlin, Heidelberg, 2011), pp. 358–373
37. V. Lakafosis et al., RF fingerprinting physical objects for anticounterfeiting applications. IEEE Trans. Microw. Theory Tech. **59**(2), 504–514 (2011). https://doi.org/10.1109/TMTT.2010.209 5030
38. D. Nouichi, M. Abdelsalam, Q. Nasir, S. Abbas, IoT devices security using RF fingerprinting Advances in Science and Engineering Technology International Conferences (ASET) 2019 (2019) 1 7 https://doi.org/10.1109/ICASET.2019.8714205
39. N. Soltanieh, Y. Norouzi, Y. Yang, N.C. Karmakar, A review of radio frequency fingerprinting techniques. IEEE J. Radio Freq. Identif. **4**(3), 222–233 (2020). https://doi.org/10.1109/JRFID. 2020.2968369
40. G. Zhang et al., Physical-layer identification of HF RFID cards based on RF fingerprinting, in *Information Security Practice and Experience* (Springer International Publishing, Cham, 2016), pp. 284–299
41. G. Khadka, M.S. Arefin, N. Chandra Karmakar, A robust detection algorithm using AC characteristics of backscatter signal for chipless RFID system (IEEE, 2019), pp. 404–407
42. J. Fan, X. Ma, L. Wu, F. Zhang, X. Yu, W. Zeng, Light gradient boosting machine: an efficient soft computing model for estimating daily reference evapotranspiration with local and external meteorological data. Agric. Water Manag. **225**, 105758 (2019). https://doi.org/10.1016/j.agwat. 2019.105758
43. F. Alzamzami, M. Hoda, A.E. Saddik, Light gradient boosting machine for general sentiment classification on short texts: a comparative evaluation. IEEE Access **8**, 101840–101858 (2020). https://doi.org/10.1109/ACCESS.2020.2997330
44. W. Liang, S. Luo, G. Zhao, H. Wu, Predicting hard rock pillar stability using GBDT, XGBoost, and LightGBM algorithms. Mathematics (Basel) **8**(5), 765 (2020). https://doi.org/10.3390/mat h8050765
45. B. Ray, R. Shah, Performance assessment of prospective PV systems in Queensland and New South Wales of Australia, in *IEEE PES GTD Grand International Conference and Exposition Asia (GTD Asia)* (2019), pp. 200–205. https://doi.org/10.1109/GTDAsia.2019.8715931

Chapter 11
Explainable Artificial Intelligence for Smart City Application: A Secure and Trusted Platform

M. Humayun Kabir, Khondokar Fida Hasan, Mohammad Kamrul Hasan, and Keyvan Ansari

Abstract Artificial Intelligence (AI) is one of the disruptive technologies that is shaping the future. It has growing applications for data-driven decisions in major smart city solutions, including transportation, education, healthcare, public governance, and power systems. At the same time, it is gaining popularity in protecting critical cyber infrastructure from cyber threats, attacks, damages, or unauthorized access. However, one of the significant issues of those traditional AI technologies (e.g., deep learning) is that the rapid progress in complexity and sophistication propelled and turned out to be uninterpretable black boxes. On many occasions, it is very challenging to understand the decision and bias to control and trust systems' unexpected or seemingly unpredictable outputs. It is acknowledged that the loss of control over interpretability of decision-making becomes a critical issue for many data-driven automated applications. But how may it affect the system's security and trustworthiness? This chapter conducts a comprehensive study of machine learning applications in cybersecurity to indicate the need for explainability to address this question. While doing that, this chapter first discusses the black-box problems of AI technologies for Cybersecurity applications in smart city-based solutions. Later, considering the new technological paradigm, Explainable Artificial Intelligence (XAI), this chapter discusses the transition from black-box to white-box. This chapter also discusses the transition requirements concerning the interpretability, transparency, understandability, and Explainability of AI-based technologies in applying different autonomous systems in smart cities. Finally, it has presented some commercial XAI

M. H. Kabir
School of Electrical and Electronic Engineering, Islamic University (IU), Kushtia, Bangladesh

K. F. Hasan (✉)
Center for Cyber Security Research and Innovation (CCSRI), RMIT University, Melbourne, VIC 3000, Australia

M. K. Hasan
School of Computer Science, Technology and Science (UITS), University of Information, Dhaka, Bangladesh

K. Ansari
School of Science, Technology, and Engineering, University of the Sunshine Coast (USC), Brisbane, Australia

© The Author(s), under exclusive license to Springer Nature Switzerland AG 2022
M. Ahmed et al. (eds.), *Explainable Artificial Intelligence for Cyber Security*,
Studies in Computational Intelligence 1025,
https://doi.org/10.1007/978-3-030-96630-0_11

241

platforms that offer explainability over traditional AI technologies before presenting future challenges and opportunities.

Keywords Explainable AI · Machine learning · Deep learning · Cyber security · Privacy · Transparency

1 Introduction

In the urbanization process worldwide, the concept of smart city gains popularity that promises to offer technology-based services. It adds merit to the traditional existed process; for example, it assists in reducing environmental footprint, improves transportation, increases digital equity, introduces new flows of revenues, improves public utilities and infrastructure, to name a few. Along with the rapid evolution of intelligent technologies and hyper-connectivity, massive amounts of data are likely to be flown through different networks requiring special attention during implementation. The dependency on data and networks makes smart cities prone to cyber-attacks more vulnerable than ever.

The rapid growth in the worldwide urban population and the increasing connectivity of this demographic make the cyber-security of smart cities more critical. This concept will become visible adopting AI with systems and devices. Most of the devices are connected with the network and some of them are utilizing the cloud perform. As a result, there is a high probability of scope to affect with cyber-attack. The explosive device is no longer considered a weapon nowadays. The malicious software named malware can shut down, interrupt, or control the smart space systems.

The scope and extent of cyberspace are increasing day by day. So it's not a human-scale problem to analyze and improve the security issues. We can able to analyze millions of events and identify malware exploiting zero-day vulnerabilities. The risk behavior leading to a phishing attack or malicious code can be easily identified with the help of AI and Machine Learning (ML) technology. The system can learn over time and utilize the past to identify new types of threats using ML. In traditional ML approaches, a data-specific model is developed from learning data. This type of model is applicable for serving particular tasks in a given environment. Due to the rapid progress in the complexity and sophistication of AI-powered systems, it becomes difficult to understand the complex working mechanisms of AI systems in general conditions. Sometimes it is very challenging when AI-based systems compute outputs that are unexpected or seemingly unpredictable. For some ML technology, the autonomy complexity and ambiguity increase, which makes the adaptation and development process interrupted. This especially holds for opaque decision-making systems, which are considered complex black-box models. To overcome these flaws, we need to transfer the classic black box models to white-box models, ensuring ML technology's interpretability, transparency, understandability, and explainability.

Explainable Artificial Intelligence (XAI) is one of the emerging branches of AI that is gaining popularity. It refers to transparent, interpretable, and understandable methods that can produce accurate and explainable models so that humans can understand the offered results [1]. The goal of XAI is to provide a way to explain the cause behind the produced outcome. Arguably, a new generation of AI approaches is emerging using XAI. XAI fundamentally establishes a relationship between the model features and outcome. XAI proposes a set of ML techniques that introduce explainable models, ensuring a high level of learning feedbacks. It maintains security privacy and trust for AI partners. Unlike the classical machine learning approaches, two different elements: explanatory module and explanation interface module, are adopted in XAI.

There is a trade-off between explainability and accuracy. Traditional models can provide better accuracy and efficiency but it has lack explainability. Introducing the XAI can offer the highest level of accuracy, efficiency and as well as explainability. Nowadays, several companies present platforms for Explainable Artificial Intelligence, including IBM, Google, Darwin AI, Flowcast, Imandra, Kybdi, and Factmata, are pioneering. According to a PwC survey reported in 2017, over 70% of the directors in the corporate world keep on the faith of utilizing the benefits of AI with their business. It also claims that AI will enhance the GDP up to $15 trillion by 2030 [2]. Thus, AI technology must become accountable and trustable are the growing demand, where XAI could be the proper choice for the AI community.

At first, this chapter discusses the black-box problems of AI technologies for Cybersecurity applications in smart city-based solutions. Later, this chapter discusses the transition and requirements from black-box to white-box to understand and ensure interpretability, transparency, understandability, and explainability of AI-based technologies in applying autonomous systems in smart cities. It presents some mention of commercial XAI platforms available to offer with traditional AI technologies before presenting XAI challenges and opportunities in the smart city domain.

2 Smart Cities: The Better Livings for Dwellers

The idea of smart city originally indicates an enhancement of quality in urban life through technology-based intelligent solutions. Nowadays, about half of the world's population lives in urban areas, and the rate of migration towards the urban areas is increasing significantly. Therefore, the dependency, efficiency, and availability of vital city services and infrastructures such as transport, energy, health care, water, and waste management are growing issues. Smart city aims to deal with all those challenges, including the capacity of city transportation, energy consumption efficiency, maintaining a green environment, improving city dwellers' economic and living standards, are a few from the list. The concept of smart cities is acquiring increasing attention on the agenda of policymakers and gaining momentum. A typical smart city is constructed with several major enabling building blocks, as shown in Fig. 1.

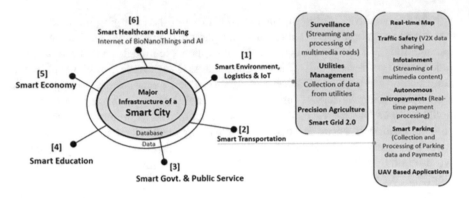

Fig. 1 Major enabling blocks of smart city

Primarily, Information and Communication Technology (ICT) equipped facilitates smart cities to automate and manage city resources and services. It participates in enhancing the city's performance and the welfare of the citizens. Smart city solutions provide intelligent network (Cellular, WiFi, Bluetooth, and Low Power Wide Area (LPWAN)) connectivity and edge processing solutions in cities to connect and improve different infrastructures.

Within the evolved concept of smart city, data can move freely within various connected services. For example, data can generate in the transportation system in response to an emergency to travel to emergency services' networks such as healthcare and government service. Similarly, public utilities (electricity, water, and gas distribution) related data can be helpful to city service operations such as taxation. Such data flows require in development of meaningful processes.

The smart city concept is now growing hand to hand with the rise of Artificial Intelligence (AI). The fundamental role of AI is to enable automated systems to increase the urban system's efficiency and productivity, reduce cost and resource consumption, improve service quality, etc. AI technologies can also serve to overcome some of the burning smart city challenges such as traffic congestion, governance, cybersecurity, and privacy. Table 1 represents some mention of AI applications in different major smart city services.

3 Security Challenges in Smart City Cyber Domain

In the smart city domain, a wide variety of sensors from extended networks work together to enforce the desired service. Therefore, smart city infrastructure is expected to consist of safe, secure, and reliable interconnected devices such as sensors and actuators to gather, process, and communicate data to enable reliable and trusted city service. However, such network heterogeneity is a great concern for

Table 1 Details of smart city building block

Services	Description of service	Notable application of AI
Smart environment, logistics, and IoT	Smart Environment, logistics, and IoT aim to provide real-time monitoring of the environment and logistics services, including infrastructure and dispersed asset monitoring. For example, different application types like water and air pollution monitoring and forest fire detection, remote and real-time monitoring of transported goods, a list of few that connect assets from anywhere to any place to make our lives safe and more manageable [3–5]	• Automated statistical information generating • Predicting building energy consumption • Business Logic generating • Environment prediction
Smart transportation	Smart transportation aims to monitor and manage transportation networks and systems to improve safety and efficiency. It is a combination of multi-functional hyper-connected intelligent transportation systems and advanced digital intelligent transportation information management. For instance, some applications monitor the vehicle load, trip scheduling, passenger entertainment, reservation, toll and ticketing systems, a list of applications alerts for overspeeding, harsh cornering, and acceleration provide a safer road [6]	• Signal control and dynamic route guidance • Traffic demand modeling • Identification of drivers based on their activities • Driver's emotion detection • Surrounding object detection

(continued)

Table 1 (continued)

Services	Description of service	Notable application of AI
Smart government and public service	Smart Govt. is a concept for the public sector that provides collection, connection, and analysis of big data generated and processed in real-time. It uses technologies to build e-government includes a Service-oriented business platform and web applications. The applications of smart government are G2C transactions, civic engagement platforms, online voting systems to stop public gatherings [7]	• Public to interact with government through the automated virtual assistant • Automatic disaster detection and management • Postdisaster reconnaissance and identification of affected areas
Smart education	Smart education is an interactive learning model with cloud-based capabilities. It is designed to enhance student engagement with teachers so that teachers can adapt to students' interests and skills. It includes data-centered intelligent education facilities for the learners. It aims to provide digital literacy, effective communication, innovative thinking to develop high-impact projects. It concludes some applications like video clavier, electronic text-books [8]	• Automated personalized learning • Differentiated and individualized learning • AI-enabled voice assistant to interact with educational tools
Smart economy	Smart economy concept combines economic features in sustainable features and eco-economic approaches like economic progress, economic prosperity, sustainable jobs. It aims to improve people's lives and adapt them to social, economic, and environmental policies. It interconnects local and global markets. It provides resource-based development to improve urban economic productivity. Smart economy can be applied to industry, tourism, maritime, mobility, payment, and banking [4]	• Monitoring economic growth and ensuring a sustainable employment rate • Increase resource efficiency and elevated competitiveness through automated sharing services, customer-tailored solutions

(continued)

Table 1 (continued)

Services	Description of service	Notable application of AI
Smart healthcare and living	Smart healthcare is a service system related to health that uses technology such as wearable Sensor networks, Body Area networks, IoT to access information dynamically and connect healthcare people, materials, and institutions. It helps in the development of telemedicine and creates personalized medical services universally. Smart healthcare is broadly used to assist medical diagnosis and treatment, health management, disease prevention, risk monitoring and management, and provide the patient's result via an automated ledger and communication services [9]	• Assist physicians with differential diagnosis of diseases • Prediction model to check up diseases by analyzing patient's electronic health records, • Providing pre-primary care information and remote monitoring to minimize the life risk

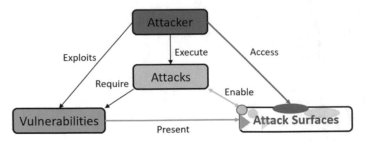

Fig. 2 Attack surface: a relation among cyber security issues [10, 11]

cyber security in the smart city realization always. As on many occasions, small low-end devices with limited capacity can participate in service-enabled communication. Due to processing and storage capacity limitations, such devices may not have robust security features such as authentication mechanisms or cryptographic ability. This leads to a possible attack surface for applications in smart city services.

Therefore, security breaches can be very real in a smart city cyber domain. An intruder can enter the system through cyber-physical system vulnerabilities. A relationship among cyber issues that leads to a cyber threat is shown in Fig. 2.

Any system vulnerability can lead to an attack surface. Since in smart cities, both high-end and low-end devices in terms of their capability connect, any Attackers can exploit low-end systems to access the service system. Moreover, any mischief can conduct some series of attacking attempts that reveal an open surface to hack the system unauthorizedly, which is shown in Fig. 2.

In the context of smart cities, cybersecurity fundamentally deals with the security of connected and internet-based services since a large part of the assets are connected to the cloud. However, it also extends the idea to protect the computer system's hardware and software system and the data flow between the networks. It makes the system steady and reliable. Cybersecurity is more effective in protecting external threats in real-time monitoring. As smart cities are the conversion of connected cities, there is a huge amount of data exchange between the entities in this domain. Cities are converged into the digital world for a sustainable economy and environment. The inhabitants of the cities generate a huge amount of data to communicate with their soundings. Smart cities have to support these excessive services and resources. Nowadays, about 55% of the world's population lives in urban areas, and it will increase up to 68% by 2050. The 2018 revision of world urbanization prospects produced by the population division of the UN department of economic and social affairs (UN DESA) notes that due to urbanization, about 2.5 billion world's population shifted to urban areas by 2050. Among them, a 90% increase in population in the metropolitan area will occur in Asia and Africa [12].

To face the needs of this large population, cities have to upgrade the traditional network with IoT technology with the adaptation of AI. This new infrastructure will be able to handle supply chains, assets, and resources management efficiently. It is challenging to handle the sensor network, analysis of data, and reasoning algorithms.

Some devices and systems that may not be cyber resilient can impose a threat to security and safety. It's quite a cumbersome job to choose a proper mitigating method that can handle threats. The security issues on IoT devices are well known. There are no security requirements/regulations imposed posed on manufacturers. So, they are not aware of the role of security vulnerability which increases system attack surfaces [13]. The role of smart cities plays a positive impact on business, city services, and people. Middle Eastern cities are now leading to provide smart cities services to their dwellers. For example, Doha is preparing to host the 2022 World Cup by adopting smart transportation and other smart services and applications. Assume that if a cyber-attack is imposed during World Cup 2022 in Doha, everything will be shut down because of a lack of sufficient safety measures [11]. Effective collaboration between vendors, device manufacturers, and governments is needed to fix a security regulation among the IoT community. Emerging standards and guidelines must be adopted to ensure system security during design, testing, and installation. Moreover, the service operator must take steps to understand the security issues and mitigation methods before the incidents occur.

4   Classic Artificial Intelligence Methods for Cyber Security in Smart City

Fundamentally, smart cities utilized Internet of Things (IoT) technologies to enable service infrastructure with remote monitoring and control capabilities. Such cyber-physical infrastructure gets technical smartness with automation capabilities by adopting Artificial Intelligence (AI). The heterogeneity of the network in smart city implementation is the greatest threat. And cyber threats can be leading to failure and devastation. The scope of cyber threats is very large. Therefore, along with traditional security measures, machine learning-based AI has also been used to enable security to those cyber-physical systems. A study of ABI research reports that 44% of the cybersecurity expenditure will spend on energy, healthcare, public safety, transportation, water, waste treatment, and other infrastructure management purpose in smart city projects [14]. Within this section's limited scope of the presentation, we will focus on some of the potential threat that deals with AI before stating fundamental challenges.

4.1   Fraud Detection

Any financial sector in smart city development is one of the indispensable parts proportionate to the urban economic growth. And this sector is probably the most targeted sector for cyber attacks naturally. Fraud in the financial sector is a common cyber-attack related to financial and reputation problems. Such attacks can be caused by data leakage and credit losses by unauthorized persons. There are several ML approaches present to detect fraud. Authors in [15] presented an SVM classifier and multi-node optimized-based fraud detection model. It used the fraudulent feature to detect the source of fraud involved in e-commerce accurately. This AI-based solution also offers to prevent further fraud. Authors in [16] proposed a Convolution Neural Network (CNN) based fraud detection model, detecting 91% of online fraud activities utilizing online transaction data. They have used the transaction data of a commercial bank. Those are a few of many examples of AI-based approaches to control measures on cyber fraud.

4.2 Intrusion Detection

Many research works are noticeable on intrusion detection mechanisms using Artificial Neural Networks, such as Deep Learning (DL) methods. Lin et al. [17] proposed a cyber intrusion detection system based on the LeNet-5 model. They have utilized over 10,000 training samples reaching an accuracy of 99.65%. Dong et al. [18] presented the Convolution Neural Network (CNN) based AE-Alexnet model utilizing the automatic encoder in real-time cyber intrusion detection system. Their model performs well in intrusion detection utilized the common KDD99 dataset achieve an accuracy of 94.32%. Khan et al. [19] improved the CNN model for intrusion detection and achieved impressive accuracy of 99.23%, outperforming the DBN model. In this space, there are many successful efforts of utilizing ML to tackle cyber issues [20].

4.3 Spam Detection

We are suffering from endless threats of spam which are unwanted emails received from various sources frequently to the users. Bosaeed, et al. proposed a system that can detect short message service spam on mobile devices. They have used Naïve Bayes (NB), Naïve Bayes Multinomial (NBM) and Support Vector Machine (SVM) classifiers. For performance analysis, they have compared the outcome of all the three classifiers, and SVM performed the best among the three classifiers [21]. Meanwhile, deep learning-based spam detection gains popularity. Chetty et al. [22] proposed a neural network-based spam detection model. The experimental results show that this model's performance decreases as the number of datasets decreases. Sharmin et al. [23] proposed another spam detection system using the Multi-Layer Perceptron (MLP), SVM, and CNN. They claim that the CNN model is suitable for detecting spam images with an accuracy of 99.02%. According to the indication of future work, they have concluded that Reinforce Neural Network (RNN) and Long Short Term Memory (LSTM) prove worthwhile in spam detection.

4.4   Malware Detection

As smart cities infrastructure is based on interconnecting networks, it is evident that malware is a significant threat that can access the network to compromise the system. Generally, malware is malicious software used to collect users' personal data and digital belongings illegally. The types and threats, complexity, and harmful nature of malware have been devastating day by day. Ransomware, crypto miners draw attention as widespread malware. Yuan et al. presented a malware classifier using Markov Images and Deep Convolution Neural Network (CNN). They have utilized two popular datasets of Microsoft and Drebin. The experimental result shows that the accuracy for Microsoft data is 99.26%, and for Drebin, datasets are 97.36% [24]. Vinayakumar et al. [25] proposed a DNN based ransomware classifier that used malware datasets collected from tweets from Twitter. They have classified the ransomware malware into 25 classes.

4.5   Traffic Analysis and Identification

Mobile devices are generating massive traffic in the IoT domain while using HTTP/HTTPS protocols, one of the prime agents of smart city realization. It is always challenging to identify and analyze any such traffic using traditional methods such as employing hostnames. It is also hard to find the signatures where devices frequently update the software versions. To tackle the traffic issue, Chen et al. [26] presented a supervised model based on CNN to tackle and identify mobile applications' traffic flow using HTTP sample data packets. This framework gets an average accuracy of 98%. Wang et al. [27] presented a classifier with an intelligent Virtualized Network Functions (VNF) selection model using Deep Neural Network (DNN) and Multi-Grained Cascade Forest. This model was successfully used to schedule the cloud network for vehicular communication to identify the transmission priority of different data packets. Lotfollahi et al. [28] proposed one-dimensional CNN for cyber traffic classifications tasks. They developed a framework based on CNN and Stacked Auto Encoder (SAE). The performance result shows a recall of 0.98 and 0.94 for recognition and traffic classification, respectively.

Overall, the detection of cyber incidences (e.g., attacks and threats) deploying traditional machine learning tools such as deep learning have been widespread in the area of cybersecurity over the last decade. Initially, it is started with conducting code analysis and deploying agents to perform signature-based detection, moving towards more self-learning and automated decompilation and extensive code analysis. It is also enabling the automated analysis and prediction of network monitoring. However, any entire AI-based automatic threat detection is becoming difficult to realize as the complexity and sophistication of the detection system create an additional problem.

Additionally, for strategic and technical reasons, AI-based tools may mislead and cannot detect actual threats. On all those occasions, it may requirer filtering out all known legit network activities to identify the threat using manual analysis. Since conventional AI-based system becomes a higher level of sophistication, it makes them a black box that drastically reduces the amount of information required to conduct manual analysis. This is the major challenge for conventional AI-based threat detection mechanisms. It remains opaque, and it is impossible to understand their back-end function and the process of inference. Therefore, explainability is required that can significantly enhance the detection capabilities with control and transparency.

5   Transition to Explainable Artificial Intelligence (XAI)

Machine learning predictive models are treated as black boxes which can be automatically trained without worrying about the domain in which they are used. This opaqueness raises many risks that are difficult to foresee during the model building process. Such as the model's declining performance due to the data drift, poor performance on the out-of-domain problems, or unfairly discriminatory behavior learned on historical data. The growing list of examples where black boxes fail spectacularly has increased interest in XAI methods. Such methods allow X-ray black-boxes models for more detailed analysis on the local or global level.

Adding interpretability to machine learning algorithms is necessary for Transportation, Smart Grid, Healthcare, Industrial applications. The general points of causality, robustness, reliability, trust, and fairness are valid in all domains. Interpretability is primarily a key to identifying causal relationships and increasing the reliability and robustness of scientific discoveries made with the help of ML algorithms. In all application domains, establishing the trust in and fairness of machine learning systems matters most in low-risk environments. In contrast, robustness and reliability are critical to high-risk environments where machines take over decisions with far-reaching consequences.

A general Machine Learning workflow is shown in Fig. 3. Firstly, data is preprocessed to train a model, and after that, the Learning function is built up using the Learning process. Then inputs are utilized to get the learned function which is used to predict the output. New inputs can be fed into the model after learning the function and the machine will return the desired prediction.

The critical thing to note in traditional ML workflow is that its prediction is without informing the reason. It can confuse the user. Users can be misled and have to believe in the machine for making the correct selection. In contrast to the traditional approach, XAI follows an advanced procedure. This procedure provides us the decision with a proper explanation about the background processes. Here the user gets additional information with output which explains the reason for prediction. An extra layer is added, which can explain the model outcome, as shown in Fig. 3. The additional layer helps to ensure impartiality in decision-making because of having an explainability module. If any biases are detected in the datasets, then they can be corrected in this layer. Moreover, explainability discovers the predictions' facts, which is beneficial

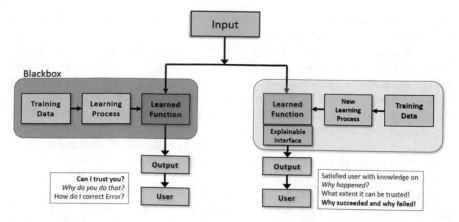

Fig. 3   A comparative view of traditional machine learning workflow (left) and explainable machine learning workflow (right)

for finding out weaknesses in outcomes. The methods that can be used to explain the behavior and predictions of trained ML models include, Partial Dependence Plots (PDP), Accumulated Local Effects (ALE), Permutation Feature Importance (PMI), Leave-One-Covariate out (LOCO), and Local Interpretable Model-agnostic Explanations (LIME).

The essence of explainability in AI tools in some major smart city enabling networks is discussed in the following sections.

5.1   Transportation System

According to the US Department of Transportation, Intelligent Transportation Systems (ITS) combines different technologies used to monitor, evaluate, and manage transportation systems to enhance efficiency, safety and cost reduction. Smart transportation becomes a reality. Several cities in the world now implement this technology. Cities like New York have adopted smart transportation to build a smart cities. They have also made a testbed for connected vehicles in Wyoming, which is a corridor for the transportation of goods. Introducing smart transportation has improved the efficiency of the supply chain and does not require the driver for the long drive. Among the several services in smart transportation systems, we will focus on smart vehicles, Intelligent Public Transport (IPT) systems, and smart city traffic management.

Smart vehicles can communicate and exchange data with other vehicles (V2V), pedestrians (V2P), or a generic network or infrastructure (V2N and V2I, respectively) [29]. The integration of IoT devices and 5G communication technologies help to make smart transportation services visible to the city inhabitants. IoT technology, together with sensors networks and embedded controllers, can be used to make any cyber-physical system manage and control remotely.

Smart self-driving cars are taking over decisions in the real world previously taken by humans and can involve severe and sometimes irreversible consequences using machine learning technology. Advanced Driver Assistance Systems (ADASs) technology provides lane-keeping and braking assistants to obtain fully autonomous driverless vehicles [30]. ML can also be an efficient tool for managing the vehicle, e.g., fuel prediction and assessing cybersecurity. IPT operators are also implementing several cybersecurity measures. However, measures are very diverse as there are neither widely accepted cybersecurity standards that align with the needs of IPT nor widely used good practices [31, 32].

In self-driving vehicles, while deploying ML technology, we should be concerned about the algorithm's complexity and be aware of the fool classifier [33]. Time complexity is an important entity in designing the safety–critical context of mobility as it mostly uses embedded technology. Moreover, a false positive result of a classifier may not be acceptable in the case of autonomous driving, such as a stop sign if the classifier treated it as a speed limit [12]. In this case, explainable AI could be a solution that can help to take a decision describing the facts behind it. Explainability could be employed as a way to disclose models' confidential data. At the same time, it also helps to ensure the reliability and robustness of autonomous driving and thus makes it safe.

Hussain et al. [34] presented a taxonomy of explainability considering fundamental questions about Why, Who, What, When, and How of explainability. They have utilized this taxonomy in human-agent systems. Scalas et al. explore [29] the role of interpretable machine learning in the ecosystem of smart vehicles. They have figured out the terms of explanations that help to design secure vehicles. In [34] the authors define explainability from an engineering view by explaining the problem of AI models. They have identified the problem of AI models in different autonomous vehicles domain for object detection, control, and perception. They proposed to introduce explainability features to the existing AI models. Soares et al. [35] proposed a new explainable self-organizing architecture and a new density-based feature selection method for the autonomous vehicle. They have focused on a self-organizing neuro-fuzzy approach to learn interpretable models automatically.

5.2 Healthcare

Proper networking between city health services using ICT infrastructure can be helpful to view the health status of community residents. Mobile devices can collect personal health data and can support private healthcare efficiently. Users can utilize a smartphone to collect health-related information using various Health apps. Moreover, users can get information about medical insurance packages, nutrition, fitness, drugs, and doctors browsing websites. The patient's observation and monitoring of disease symptoms and behavioral patterns (activity, sleep disorder, mobility pattern, phone conversation) become easy using the flavor of IoT. The mobile healthcare system can detect the changes of different patient symptoms and take the safety measure accordingly both online and remotely [36]. Mobile devices can play an important role, from individual monitoring to community-level monitoring. A medical practitioner can keep important record notes on diagnosis for a patient. Severe and chronically ill people can utilize social media to share their illness, remedy experiences, and support the community. Some users may not prefer physical support; in that case, online support using mobile devices and IoT is helpful.

People who are suffering from long or short-term chronic diseases can be tracked using mobile ICT. Users can also keep a record of their drug-taking schedule and can set alarms as a reminder. Awareness can be patronized among the inhabitants as prevention. Online medical technologies are getting popular among city dwellers using mobile devices and apps. Moreover, the integration of ambient sensors into smart city networks makes it possible to provide services to vulnerable and non-tech-savvy patients. City designers and researchers urge for integrating smart ICT and Healthcare. The advancement in pervasive computing and machine learning accelerates the integration of smart healthcare in smart city technology. Now vast amounts of personal health data are shared; as a result, privacy issues arise. Although healthcare system utilizing Artificial Intelligence (AI) has shown, it's competency in providing cost-effective and human-centric application. The performance of medical diagnosis, image-video processing, resource tracking can be improved by introducing the power of AI. Therefore, the convergence of AI technologies to healthcare is a demand of current application in the smart cities domain.

Recent advancement in AI technology demonstrates remarkable performance in implementing AI and machine learning in healthcare. Healthcare requires a high level of accountability and transparency. Security and privacy are two main issues that are the barriers to accommodating technology in healthcare. The current ML solutions are mostly dark in nature based on the black-box model, which doesn't become explainable. It is against the moral responsibility of physicians to use the black-box medical decision-making system [37]. This system can not provide the explanation behind the reasoning. In the case of consulting an expert AI system, the physician may ask which factors are considered of reasoning. The traditional AI system, which is opaque in nature, can not provide the background behind the reasoning outcome to the physician. In order to provide the answer of the physician, the expert system has to build the knowledge of explanation that satisfies the physician. Thus explanation provides reliability, improves transparency in the decision-making system.

In [9], authors have presented suggestions for explainable medical AI systems. They explored the idea of explanation in view of human-centric orientation. Meacham et al. [38] have designed and developed a web system to predict and explain patient readmittance using machines. They have utilized a logistic regression-based ML algorithm with an explainability function and a web interface corresponding users' interfaces.

5.3   Smart Grid

The increasing population in cities causes a decrease in fossil energy sources. So renewable energy sources are the best solution for supporting the energy demand. About 75–80% of the energy of the world is consumed in cities. This energy consumption generates about 80% of the world's greenhouse gas emissions [7]. Smart Grid (SG) is the key item of energy infrastructure for sustainable smart cities. It aims to reform fossil energy resources into renewable energy resources. SG plays a critical role in reducing environmental pollution due to the use of non-renewable energy resources.

Moreover, it can provide lower energy costs and reliable energy service. The traditional non-smart grid lacks online monitoring and controlling facilities which is a flaw for a real-time solution. The smart city energy system should have facilities to connect each energy system and grid for real-time remote control and monitoring. This ensures efficient energy generation and consumption. A prediction shows that the 15.8% smart energy market will invest around US$248.36 billion for producing smart energy systems [39]. In a review paper [40], the smart plug technology is illustrated, potentially contributing to green energy, SG/microgrids, and smart cities. In [41], the authors presented current state-of-the-art techniques used to predict advanced intelligent loads in SG.

Smart meters provide monitoring of the exact and real-time electricity usage, offering bill information adopting Advanced Metering Infrastructure (AMI). It also helps to insist on carbon emissions and assists in making intelligent decisions. Smart grids support Electric Vehicles (EVs) to detect and accept the produced/stored energy from the users' side. Smart grid offers smart load and building automation, leading to increased energy efficiency, safety, and comfort. The smart city project of Malaga, Spain, launched in 2009, aims to be the largest energy-smart city [42]. It utilized smart meters for remote management to improve energy efficiency. A light-emitting diode (LED) network was also built up for street lighting. Another project of future smart cities started in 2015 at the EU named Europe's future smart cities. Eindhoven, Manchester, and Stavanger, in the Netherlands, UK, and Norway, respectively, are equipped with LED networks and serve as testbeds. The plan is to imply the findings to the other three cities Leipzig, Sabadell, and Prague, in Germany, Spain, and Czech Republic, respectively [38]. Several technologies related to energy, carbon emission, and air quality are imposed on this area, and the performance is closely observed. The main focus of this project is to update the SGs by integrating real-time monitoring, automation, and self-controlling facilities.

AI approaches have been applied to various smart grid applications, such as demand response, predictive maintenance, and load forecasting. However, traditional AI techniques are considered a "black-box" due to their lack of explainability and transparency. AI-based methods are utilized to predict the energy consumption aims to save energy [43]. Consider the case; a heating device could be turned off if the device fails to operate on the low power source. However, the heating device can also be turned off to save energy when the desired ambient temperature is reached or when the predicted energy consumption is reached above the threshold. This scenario can be modeled using XAI model only, which can explain the real fact why the heating device is turned off. In traditional black-box AI techniques, it is not possible to know the fact behind the reasoning. XAI could be an interesting research idea because it connects with the expert to the finding of the AI system by explaining the cause behind the reasoning.

In [44], the authors proposed an approach to introduce XAI in ontologies and semantic technology to provide autonomous car decision-making and energy efficiency solutions. Kuzlu et al. [11] utilize three XAI tools (LIME, SHAP, and ELI5) [33] to produce high-level explanations about the outcome of the system in smart grid applications.

6 Commercial XAI Platforms

XAI technology is emerging. Several commercial platforms are formed to support this increasing demand. In this section, we will look at some of the popular XAI platforms in brief.

IBM. IBM is one of the pioneers of developing XAI platforms which has opened an advanced cloud-based AI tool using XAI technology. It aims to expose the causes behind the present state of art AI recommendations. They conducted an extensive survey that reported that over 60% of its executives feel uncomfortable with the AI solution's traditional black-box approach. In other ways, such a survey inspires them to propel the explainability in their AI solutions [45].

IBM ensures fairness, robustness, explainability, accountability, and value alignment in AI applications. IBM aimed to utilize these virtues in all respects to the whole lifecycle of their AI application. IBM integrated XAI frameworks and toolsets into the IBM Cloud Paks, which supports the data platform. They use AIx 360 toolkit, which includes a set of algorithms that cover several domains of explanations. This tool is made for algorithmic research for different practices of domains. As a result, their AI technology supports businesses in a controlled, safe and measurable way. The AI practitioner can build, run, and manage AI models and can optimize decisions anywhere on IBM Cloud named Paks® using IBM Watson® Studio for Data [46]. The application area of IBM using XAI is Healthcare, Financial Services, and Education. IBM Cloud Paks can increase the model accuracy by 15–30%.

Google. Another XAI-enabled AI platform has been launched by tech giant Google, which explains the performance of several of Google's features. The feature attributions are generated using AutoML Tables and AI Platform for model predictions in XAI solution platform. Whereas the model behavior can be investigated visually using the What-If Tool. The AI practitioner, such as data scientists, can explore to improve the usable datasets or models. The model performance is also debugged using AI Explanations in AutoML Tables, AI platform predictions, and AI platform notebooks provided by Google.

Moreover, the model behavior can be investigated using the What-If Tool minutely. The ultimate target is to build users' trust and improve the transparency and fairness of machine learning models through human-interpretable explanations. We can predict deploying an AutoML table or AI platform, providing a real-time score [47]. The ground truth of predictions input can be sampled using continuous evaluation capability from the trained ML models. Mainly the XAI-enabled platform designed by google explains the image recognition and code sampling features of google. Arguably, Google has a pervasive influence on smart city realization from an intelligent car to smart home development.

Darwin AI. Drawin AI, an XAI service provider company, enables enterprises to structure AI on which they can stand. Darwin AI is active in building a process known as Generative synthesis. It aims to make the explainability real, granting the makers to realize the interior functions of deep neural networks. Within this framework, Darwin AI employs an artificial neural network-based deep learning method. It offers an Explainability toolkit and network diagnostics features. Moreover, it also provides some other network automation for connectivity applications. It focuses on detecting defects and adaptive factory automation, one of the significant smart infrastructure sectors.

Flowcast. Flowcast is an API-based solution that targets bringing out the black box models by fusing different company systems. Flowcast clarifies the relationships between input and output values of different patterns to establish explainability. It utilizes its own AI technologies to develop credit assessment models with the explainable module. Such an enhancement comes up with the potential to transform credit regulations completely into the new ability of explainability. Among a selection of ML algorithms used by Flowcast, the boosted trees algorithm is an opaque variant; they had to find a way to explain system assessments clearly. Thus, explainability uses SHAP along with NLP to provide plaintext phases presenting outputs in layman's terms. Therefore, help financial institutions adhere to regulatory compliance by reducing risk and unlocking credit for all.

Imandra. Imandra is an automated reasoning technology that is a digitization and AI for capital market APIs. Imandra democratizes automated reasoning to enable algorithms to be explainable, trustworthy. They present the Reasoning as a Service (RAAS) feature, which fetches these methods nearer to those except a specialized background in these areas. Initially, Imandra started with the financial sector and has worked with major investment players. It has focused on service designing, testing, and conducting AI-based audits of its complex trading systems, which has a great future in smart city applications.

Kyndi. Kyndi is a pioneering text analytics company that utilizes an explainable natural language processing platform. It optimizes human cognitive performance by transforming business processes through auditable AI systems. Kyndi provides the auditable trail of reasoning if the stakeholders of organizations seek an explanation of the output. Kyndi uses a well-established programming language called Prolog with RPA tools to build bots that automate fact, inference, and concept generation in almost any vertical. Its expanded model aids in encoding the semantics of documents, repositories, and domains in query-able knowledge graphs. Kyndi aims to utilize its tools for smart city application in association with different stakeholders, including federal government sectors, such as Defense, Intelligence, finance, healthcare, IT, and infrastructure.

Factmata. Fake news is created maliciously to misinform people by generating more traffic to sites, maybe publishing scandalous headlines or politically biased content focusing on chaos. Content Moderation performed by humans is time-consuming, repetitive, and open to errors. Factmata is an AI solution that makes AI power tools detect problematic content from inter. Factmata is fighting against online fake news with XAI techniques to justify segregation [48].

7 Opportunities and Challenges of XAI

The Explainability model could significantly increase the detection competency, and the prediction result could be trustworthy. Many of the application domains are handling personal and legal issues data. Hence aware should be taken to find the cause and issues behind the outcome of reasoning provided by the trained model. The model involved in explainability should not generate a large set of negative results. Much concentration should be focused on explaining the trained model in prediction and classification applications. In open-source intelligence, explainability could provide a means for detecting attacks, impacting mitigating their influence [49].

Nowadays, ML models used in cyber security do not have an Explanations module. It is very important to find the cause behind reasoning because system administrators can demand more information other than informing only the binary decision. A realistic threat model can be utilized to know about the strength of mitigation methods in the cyber-security domain. There are several XAI methods available nowadays. The good part is that these XAI methods can expose the mode of processing to produce the decision of trained ML models to some extent. Explainability would also have a significant impact on Adversarial Machine Learning as well.

However, several research challenges exist to develop methods for explainability from the technical, legal, and practical points of view. Technical challenges not only deal with XAI system development but also evaluation and interpretation. It is because the XAI system must include an explicit representation of the component parts to support the appropriate and systematic interpretation of working functions. In such cases, a set of semantics representations can help construct an explanation that a human receiver understands. It is, however, may still tricky for a human to interpret the functioning of the XAI methods. Therefore, XAI systems should aim to deal with the role of the learning method. It should have a set of open procedures to reason the inference process. Overall, explanations of the AI model are needed for the proper design and functioning of the correct decision-making system. Although, they are not the ultimate goal of designing AI system. XAI applies knowledge and expertise to design a sensible and transparent AI system.

The application of XAI methods in any specific domain, on the other hand, can face the question of legitimacy. This can happen when handling data involving legal issues, decisions, life, and finance, gathering more intelligence on the people involved. The explainability module in a learning model utilizes to infer data about entities. These allowed can produce significant data, underlying or hidden correlations, and causality is a challenge.

The success of XAI depends on fruitful human–machine interaction. It is considered that both machines' outcome human decision would be consistent irrespective of the ground truths. However, there is high uncertainty in the ground truth and cannot be fully explained. Current XAI methods and techniques predominantly depend on input–relevant parts and do not incorporate the human model's notion [50]. Hence the concept of causability arises, which is a challenge for XAI realization.

8 Conclusion

This chapter presented an overall analysis of the prospect of explainable artificial intelligence in the cyber security domain of smart city applications. We conducted our study to portrait the opportunity and necessity of XAI using secondary data. Undoubtedly, the advancement of AI is primarily dependent on the flourishment of ML techniques. It will pave the path to handle the high-frequency, real-time application for the smart city. It is, however, difficult to see how such methods work in underlining. This is very real to combat any cyber security issues. If any AI-based cyber security component does not have forensic value but remains as a black box, it certainly has less credibility than all. Therefore, XAI has a great role in ensuring smart city deployment's security, privacy, and trustworthiness.

References

1. D.V. Carvalho, E.M. Pereira, J.S.J.E. Cardoso, Machine learning interpretability: a survey on methods and metrics, 8(8), 832 (2019)
2. D.A.S.R.A.G. Verweij, Sizing the prize: What's the real value of AI for your business and how can you capitalise?" https://www.pwc.com/gx/en/issues/analytics/assets/pwc-ai-analysis-sizing-the-prize-report.pdf
3. M.H. Kabir, M.R. Hoque, S.-H.J.I.J.O.S.H. Yang, Development of a smart home context-aware application: a machine learning based approach, 9(1), 217–226 (2015)
4. J.-L. Briaud et al., Realtime monitoring of bridge scour using remote monitoring technology, Texas Transportation Institute (2011)
5. K.F. Hasan, Y. Feng, Y.-C. Tian, GNSS time synchronization in vehicular ad-hoc networks: benefits and feasibility. IEEE Trans. Intell. Transp. Syst. 19(12), 3915–3924 (2018)

6. C.B.Y. Zhang, Y. Zhang, J. Zhang, J. Xu, Automatic mobile application traffic identification by convolutional neural networks, in *IEEE Trustcom/BigDataSE/ISPA* (2016), pp. 301–307. https://doi.org/10.1109/TrustCom.2016.0077

7. T. Nam, T.A. Pardo, Smart city as urban innovation: focusing on management, policy, and context, in *Proceedings of the 5th International Conference on Theory and Practice of Electronic Governance* (2011), pp. 185–194

8. Smart City Malaga, https://malagasmart.malaga.eu/en/sustainable-and-safe-habitat/energy/smartcity-malaga/. Accessed 7 Sept 2021

9. Y. Xie, G. Gao, X.A. Chen, Outlining the design space of explainable intelligent systems for medical diagnosis (2019)

10. K. Fida Hasan, A. Overall, K. Ansari, G. Ramachandran, R. Jurdak, Security, privacy and trust: cognitive Internet of vehicles (2021), arXiv e-prints arXiv:2104.12878

11. M. Kuzlu, U. Cali, V. Sharma, Ö.J.I.A. Güler, Gaining insight into solar photovoltaic power generation forecasting utilizing explainable artificial intelligence tools, **8**, 187814–187823 (2020)

12. UN DESA, The revision of the world urbanization prospects, population division of the United Nations. https://www.un.org/development/desa/en/news/population/2018-revision-of-world-urbanization-prospects.html

13. I. Butun, P. Österberg, H.J.I.C.S. Song, Tutorials, Security of the Internet of Things: vulnerabilities, attacks, and countermeasures, **22**(1), 616–644 (2019)

14. J. Wilson, N.J.I.T.O.M.C. Patwari, Radio tomographic imaging with wireless networks, **9**(5), 621–632 (2010)

15. A.S. Sadiq, H. Faris, A.-Z. Ala'M, S. Mirjalili, K.Z. Ghafoor, Fraud detection model based on multi-verse features extraction approach for smart city applications, in *Smart Cities Cybersecurity and Privacy* (Elsevier, 2019), pp. 241–251

16. Z. Zhang, X. Zhou, X. Zhang, L. Wang, P.J.S. Wang, C. Networks, A model based on convolutional neural network for online transaction fraud detection, **2018** (2018)

17. W.-H. Lin, H.-C. Lin, P. Wang, B.-H. Wu, J.-Y. Tsai, Using convolutional neural networks to network intrusion detection for cyber threats, in *2018 IEEE International Conference on Applied System Invention (ICASI)* (IEEE, 2018), pp. 1107–1110

18. Y. Dong, R. Wang, J. He, Real-time network intrusion detection system based on deep learning, in *2019 IEEE 10th International Conference on Software Engineering and Service Science (ICSESS)* (IEEE, 2019), pp. 1–4

19. R.U. Khan, X. Zhang, M. Alazab, R. Kumar, An improved convolutional neural network model for intrusion detection in networks, in *2019 Cybersecurity and Cyberforensics Conference (CCC)* (IEEE, 2019), pp. 74–77

20. M.Z. Alom, T.M. Taha, Network intrusion detection for cyber security on neuromorphic computing system, in *2017 International Joint Conference on Neural Networks (IJCNN)* (IEEE, 2017), pp. 3830–3837

21. S. Bosaeed, I. Katib, R. Mehmood, A fog-augmented machine learning based SMS spam detection and classification system, in *2020 Fifth International Conference on Fog and Mobile Edge Computing (FMEC)* (IEEE, 2020), pp. 325–330

22. G. Chetty, H. Bui, M. White, Deep learning based spam detection system, in *2019 International Conference on Machine Learning and Data Engineering (iCMLDE)* (IEEE, 2019), pp. 91–96

23. T. Sharmin, F. Di Troia, K. Potika, M. Stamp, Convolutional neural networks for image spam detection, **29**(3), 103–117 (2020)

24. S. Yang et al., Scalable digital neuromorphic architecture for large-scale biophysically meaningful neural network with multi-compartment neurons. J. IEEE Trans. Neural Netw. Learn. Syst. **31**(1), 148–162 (2019)

25. R. Vinayakumar, M. Alazab, K. Soman, P. Poornachandran, A. Al-Nemrat, S. Venkatraman, Deep learning approach for intelligent intrusion detection system, **7**, 41525–41550 (2019)

26. Z. Chen, B. Yu, Y. Zhang, J. Zhang, J. Xu, Automatic mobile application traffic identification by convolutional neural networks, in *2016 IEEE Trustcom/BigDataSE/ISPA* (IEEE, 2016), pp. 301–307

27. J. Wang, B. He, J. Wang, T. Li, Intelligent VNFs selection based on traffic identification in vehicular cloud networks, **68**(5), 4140–4147 (2018)
28. M. Lotfollahi, M.J. Siavoshani, R.S.H. Zade, M. Saberian, Deep packet: a novel approach for encrypted traffic classification using deep learning. J. Soft Comput. **24**(3), 1999–2012 (2020)
29. M. Scalas, G. Giacinto, On the role of explainable machine learning for secure smart vehicles, in *2020 AEIT International Conference of Electrical and Electronic Technologies for Automotive (AEIT AUTOMOTIVE)* (IEEE, 2020), pp. 1–6
30. N. Dasanayaka, K.F. Hasan, C. Wang, Y. Feng, Enhancing vulnerable road user safety: a survey of existing practices and consideration for using mobile devices for V2X connections (2020), arXiv preprint arXiv:2010.15502.
31. K.F. Hasan, C. Wang, Y. Feng, Y.-C. Tian, Time synchronization in vehicular ad-hoc networks: a survey on theory and practice. Vehic. Commun. **14**, 39–51 (2018)
32. K.F. Hasan, T. Kaur, M.M. Hasan, Y. Feng, Cognitive internet of vehicles: motivation, layered architecture and security issues, in *2019 International Conference on Sustainable Technologies for Industry 4.0 (STI)* (IEEE, 2019), pp. 1–6
33. T. Gu, B. Dolan-Gavitt, S. Garg, Badnets: Identifying vulnerabilities in the machine learning model supply chain (2017)
34. F. Hussain, R. Hussain, E. Hossain, Explainable Artificial Intelligence (XAI): an engineering perspective (2021)
35. E. Soares, P. Angelov, D. Filev, B. Costa, M. Castro, S. Nageshrao, Explainable density-based approach for self-driving actions classification, in *2019 18th IEEE International Conference On Machine Learning And Applications (ICMLA)* (IEEE, 2019), pp. 469–474
36. D.J. Cook, G. Duncan, G. Sprint, R.L. Fritz, Using smart city technology to make healthcare smarter, **106**(4), 708–722 (2018)
37. A.J. London, Artificial intelligence and black-box medical decisions: accuracy versus explainability. J. Hastings Center Rep. **49**(1), 15–21 (2019)
38. S. Meacham, G. Isaac, D. Nauck, B. Virginas, Towards explainable AI: design and development for explanation of machine learning predictions for a patient readmittance medical application, in Intelligent Computing-Proceedings of the Computing Conference (Springer, 2019), pp. 939–955
39. A. Gaviano, K. Weber, C. Dirmeier, Challenges and integration of PV and wind energy facilities from a smart grid point of view, **25**, 118–125 (2012)
40. N.K. Suryadevara, G.R. Biswal, Smart plugs: paradigms and applications in the smart city-and-smart grid. J. Energies **12**(10), 1957 (2019)
41. S.N. Fallah, R.C. Deo, M. Shojafar, M. Conti, S.J.E. Shamshirband, Computational intelligence approaches for energy load forecasting in smart energy management grids: state of the art, future challenges, and research directions, **11**(3), 596 (2018)
42. J.L.Y. Wang, Y. Chen, M. Gruteser, J. Yang, H. Liu, E-eyes: device-free location-oriented activity identification using fine-grained WiFi signatures, in *Proceedings of Annual International Conference Mobile Computing Networking, MOBICOM* (2014), pp 617–628. https://doi.org/10.1145/2639108.2639143
43. N. Petrović, D. Kocić, Data-driven framework for energy-efficient smart cities, **17**(1), 41–63 (2020)
44. N. Petrović, M. Tošić, Explainable artificial intelligence and reasoning in smart cities, YuInfo 2020, pp. 1–6 (2020)
45. M. De Sanctis, E. Cianca, S. Di Domenico, D. Provenziani, G. Bianchi, M. Ruggieri, Wibecam: device free human activity recognition through wifi beacon-enabled camera, in *Proceedings of the 2nd workshop on Workshop on Physical Analytics* (2015), pp. 7–12
46. IBM, Explainable AI, [online] Available at: https://www.ibm.com/se-en/watson/explainable-ai (Accessed 4 Aug 2021)
47. G. Cloud, Explainable AI, [online] Available at: https://cloud.google.com/explainable-ai (Accessed 3 March 2022)
48. Factmata, Helping organizations understand [online] Available at: https://www.factmata.com/ (Accessed 3 March 2022)

49. S.M. Devine, N.D. Bastian, Intelligent systems design for malware classification under adversarial conditions (2019), arXiv preprint arXiv:1907.03149
50. A. Holzinger, G. Langs, H. Denk, K. Zatloukal, H. Müller, Causability and explainability of artificial intelligence in medicine. Wiley Interdisc. Rev.: Data Min. Knowl. Discov. **9**(4), e1312 (2019)

Chapter 12
Explainable Artificial Intelligence in Sustainable Smart Healthcare

Mohiuddin Ahmed and Shahrin Zubair

Abstract Artificial Intelligence (AI) is the capability of a system to execute tasks similar to decisions taken by human intelligence. AI has been certainly the hotspot for Internet of Health Things (IoHT) and has brought revolutionary changes in the health community. But yet the healthcare providers and the researchers' demand for explanation of the resulting predictions made by the system on the basis of the health data trained in the Machine Learning model was not satisfied. Thus, the field of Explainable Artificial Intelligence (XAI) has been explored by the researcher community to provide explanation to the predictions made by the machines and ensure accuracy in the absolute healthcare infrastructure. Since blindly relying on the decisions made by the machine for saving a human soul without proper understanding of the underlying logic is inappropriate, in this condition XAI assists the medical care team to understand the logic and counter check the decisions before implementing on the patient for a better cause. Our aim is to highlight the reasons of adopting XAI in the healthcare domain in this book chapter and discuss the basic concept behind it on how it can contribute towards reliant AI-based solutions to the healthcare.

Keywords Artificial intelligence (AI) · Machine learning · Cybersecurity · Healthcare system · Explainable artificial intelligence (XAI) · Internet of health things (IoHT) · Human intelligence · Cyber-attacks · Big data

1 Introduction

Artificial Intelligence (AI) is a set of models and methodologies that are used to extract knowledge from a collection of data [1]. From healthcare to criminal justice, it progressively aids in carrying consequential human decisions. AI in general and

M. Ahmed (✉)
School of Science, Edith Cowan University, Perth, Australia
e-mail: m.ahmed.au@ieee.org; mohiuddin.ahmed@ecu.edu.au

S. Zubair
Department of Computer Science and Engineering, International Islamic University of Chittagong, Chittagong 4318, Bangladesh

© The Author(s), under exclusive license to Springer Nature Switzerland AG 2022
M. Ahmed et al. (eds.), *Explainable Artificial Intelligence for Cyber Security,*
Studies in Computational Intelligence 1025,
https://doi.org/10.1007/978-3-030-96630-0_12

machine learning (ML) in particular have remarkably proven to have a disruptive socio-economic impact. These technologies evidently play promising roles in introducing intelligentization of clinical practices. Healthcare and medicine is at the cutting edge of a revolutionary and instantaneous transition towards data dependency. The rise in data acquisition and networked system technologies is becoming the significant driver behind the evolution of techniques for management and analysis of biological data. Such situations should be considered as an exceptional room for both AI and ML along with several related approaches for knowledge extraction particularly from data. As a matter of fact, the linkage between AI and healthcare share a deep-rooted past yet the adoption of AI in clinical settings is still bounded due to trust and transparency issues [2] . Despite the current AI systems are to an extent competent enough to tackle real-world problems, but the black-box nature of these systems in few of the cases is creating hindrance in the mass acceptance of AI. A 'black box' in general refers to a system or program that grants to identify the input and output, but gives no visible portrait of the ongoing processes and workflow with them. ML is vitally being chosen as a component to combat the unfolded cyber threats but its effectiveness is yet limited due to its inability to clear up and simplify its thoughts and actions to the users. One of the major drivers for succesfully adopting these techniques is to ensure how accurately the domain experts and the human users are capable of understanding and trusting their performance. The clarity in the output of ML in providing cybersecurity is crucial as here these experts need far more information than a simple binary output for their analysis. These challenges have encouraged research interest in the Explainable AI commonly known as XAI [3]. It can majorly contribute in developing techniques to make the systems "intelligible" or "explainable". XAI is that domain where several techniques are evolved to explain and interpret predictions made by these modern AI systems. It can be used to analyze and diagnose the health data by AI-based systems to accomplish transparency, accurate tracing of result, accountability in the healthcare sector. It can efficiently make the results of these AI systems more understandable and clear to the human users. The invincible penetration of AI and ML across health industry with its vital impact on critical decision-making processes but without detailed information about the specific reasons to take such decisions leads to uncertainty. Therefore, XAI is one of the new AI techniques that is capable of making decisions explainable and understandable. ML as mentioned earlier is growing popularity in cybersecurity applications such as in an intrusion detection system (IDS) but most of the ML models are considered to be a black-box. XAI is a promising solution to interpret these ML models to escalate trust management, transparency, and visibility by granting the domain experts and the users to prominently understand the underlying data evidence and causal reasoning [1]. And in cyber security applications such as IDS, it is significant to understand and being able to interpret the affluence of the malicious data to detect any unauthorised intrusion in the system. This chapter aims to highlight the impact of XAI in a broader aspect relating it to the present condition of the cyber-attacks held on Internet of Health Things (IoHT).

2 Why XAI is Important?

The expansion of AI is having a substantial impact on every fraternity. Almost every common activity held in daily lives involves AI for making decisions. Taking the basic instants, such as movie or ad recommendations based on an individual's choice on each and every social media platform is very much familiar [1]. It does not only limits to recommendations but also involves predicting or taking necessary actions accordingly. But AI still considerably lags behind while taking critical and powerful decisions for any life-changing scenario. despite having effective result and predictions [1], they are yet considered to not being able to provide transparency in their algorithms used, especially in cases where Machine Learning (ML) is used. It becomes challenging to get a comprehensive portrayal of the internal procedure of the work. This ultimately leads to ambiguous situations where the system becomes incompetent to explain the internal procedures and the reasoning behind taking major and risky decisions of life [1]. Relying on a system that cannot explicitly explain the purpose of making a decision leaves question on the effectiveness of the system. Thus to fill the void, researchers have proposed Explainable AI (XAI) [1] to ensure fair, vague and accountable decision making by the system. The prime focus of introducing XAI [1] widely is to present several methods for generating more highly effective interpretable models. It is significant to have XAI as the core technology when the user demands a system to be able to appropriately manage its results in a transparent, understandable and efficient manner. The study on the literature review of XAI [1] shows that there are four valid reasonings behind the need of having XAI as a reliable medium for making or predicting actions based on accurate decision. Discovering, improving, justifying and controlling is the fundamental basis on which explainability is significant in the algorithms. The ability to discover new facts for information gathering and being able to improve a model based on understanding the flaws as well as controlling it in crucial situations with accurate justification is the prominent advantage of XAI [1]. But there are certain situations where explanation is not required for every small matter in order to prevent less effective system that is biased towards providing explanation with variety of results. Thus, it is crucial to get an insight on why and in which scenarios do we need explanations in AI systems. As for example in any situations where the level of non-transparency is low and encountering error-based problems is relatively negligible [1], the explainability can be sidestepped as providing explanation in an AI system is unquestionably expensive. Thus, XAI is important in potential decision making where a single fault while making predictions or taking decisions can lead to dreadful conditions accordingly.

3 XAI in Healthcare

The amalgamation of the technologies [2] such as, AI, Cloud computing, and Internet of Things together contributes towards building a Smart healthcare for enabling an effective, ready-to-go, and reliable healthcare system. Such system allows healthcare applications and wearable gadgets to keep a trace of patient's real-time health data and monitor accordingly facilitating self-control for maintaining good health. AI in such a system proactively promotes regular medical checkup, initial clinical diagnosis of sickness, and planning for needed medical therapies and treatment for the suspected disease. The ethical issues [2] relating to trust the AI based smart systems in the healthcare field arises the need for introducing AI models that explains every prediction or decision that has been made by it. Thus, this requires the XAI methods for providing explanations to such predictions or decisions taken for the patients' by the system to treat them according to their health demand. The basic need and advantages for the collaboration of XAI methods in the smart healthcare system is to elevate transparency for boosting trust within the system, trace the circumstances of the failed outcomes and to assist in improving the models. There have been a number of scenarios where several solutions based on XAI have been proposed and applied in the healthcare sector. There are few AI models [2] that are self-explainable but yet have challenges associated with them in order to achieve the highest level of accuracy in the desired outcome. The massive ongoing research in XAI addresses the dire need of developing infrastructures and models to aid in explaining and understanding the predictions and decisions made by the AI based systems. The paper [2] proposed an idea to utilise the existing models of XAI simultaneously with medical knowledge. The motive behind proposing it was to procure and make the most of AI-based healthcare systems. Here, the real-time health data is obtained from the patient by a healthcare application. The trained models of AI are being used to predict the possibilities of certain diseases and anomalies in the patient. The XAI methods now use these both relevant data, i.e. patient's health data and predicted results to initiate explanations accordingly. The specialists [2] can assist the system by analyzing the generated explanations with their attained knowledge to approve and validate the AI-based predictions for enabling transparency in the procedure. Based on whether or not the predictions are correct, the analysed explanations can be further utilised for either generating health advice or trace elements responsible for erroneous predictions and improve the associated AI model. The XAI in the clinical domain [4] must necessarily need to consider the fact that multiple varied data contributes to a single relevant outcome. Thus, it is significant for the physicians to appropriately being able to interpret and trust the decisions made by the machines and it can only be possible when the used algorithms are transparent. The research in the field of XAI based smart medical system [4] shows that there is a considerable need for a high performance in training for several distinctive machine learning techniques and human-computer interaction methods. There remains an innate pressure [4] between the accuracy in predicted decisions attained by the machine learning algorithms and understanding them. There are often situations where few methods with high performance rate has minimal transparency whereas other methods with appropriate reasoning and explanation are erroneous. It is crucial for the machine learning algorithms [5] utilised in the health domain to explain the accurate reasoning behind taking a life-changing decision for a patient. Since now-a-days the applications of machine learning are highly integrated into the healthcare domain for serving the purpose of patient care, providing explanation based on the solutions to the clinical specialists is essential. These explanations [5] can further assist these specialists in various sectors.

4 Applications of Artificial Intelligence in Healthcare

The practise of utilizing smartphones and relevant devices for connecting and delivering remote medical health services to the needful is referred to as Mobile health (m-health) [3].

Today it is often considered to be the significant development of science and technology. It successfully involves Artificial Intelligence (AI) [3] for presenting an efficient and operative healthcare model for the society. The m-health based applications are categorized into three parts [3]: fitness-based primary-care, medicine-based information, and for administering the overall healthcare. m-Health [3] is highly contributing in expanding remote health monitoring services and enhancing the quality of care and safety provided to the patients. m-health allows to collect data from several points simultaneously and grants distant medical assistance in a timely manner. These gathered data can be further used by the communities to interpret the conventional trends and granting them to revise the healthcare quality. AI [3] is the procedure to exhibit machine intelligence stimulated by the natural-born human intelligence. Machine learning (ML) is an area of application of AI that allows to generate opportunities to the system to automatically train and learn from the prior experiences with the data without the need to be explicitly programmed. The implementation of AI has been approached in various healthcare sectors [3] such as, cancer research, diabetes, psychological health, cardiology, recognizing Alzheimer and cardio-respiratory disease, and several studies related to strokes, etc. as shown in Fig. 1. Larburu et al. [6] has proposed an AI-based m-Health application to aid the patients in avoiding cardiac arrest. There have been numerous circumstances in the

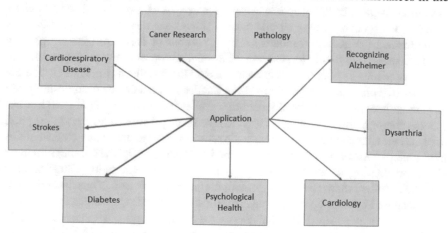

Fig. 1 Applications of AI in healthcare

present market of such applications where they have been generating false alarms claiming to identify the symptoms of cardiac-arrest. But the proposed model has successfully been able to decline the rate of false alert and accurately predicts the potential danger of cardiac arrest in the patient. They [6] have applied Naive Bayes Classifier for acquiring the finest prediction in the criteria but failed to precisely detect the risk factors in the patients who have previously undergone any cardiac surgeries. Another study [7] presented a context-sensitive AI-based system that senses the state and behavioral pattern of a human being in depression and provides them the back to change those patterns by positively reinforcing the changes needed to have a better mental health. They have used decision trees as well as regression in combination with the sensor-based mobile devices. The accuracy rate was excellent in predicting the behavioral patterns but they lagged behind in providing promising results for every natural mood and emotions which certainly needs to be upgraded. The study [8] had proposed speech recognition automated system for people suffering from dsyarthria. It also aids in generating voice message. The hidden Markov models were utilised for every word that is voiced to a speech model for deciding their general closeness in respect to each person. Its methodology claims to have an accuracy of only 67% with a disadvantage of having less accuracy rate in the hardware used for speech recognition. AI has exceptionally made advancements in the field of image classifications [9] which has been assisting the dermatological diagnosis to reach the level of accuracy. It assists the dermatologists to detect and identify skin lesions such as melanoma via computer algorithms. The cases of skin cancer is rapidly increasing in the western part of the world and AI has the highest probability to aid in diagnosing as well as providing both primary and secondary healthcare [9]. It is expected that AI can be the key tool in bringing revolutionary changes in the dermatological and other vital sectors of the healthcare system with its proper ethical implementation. Inspection [3] is a key factor in treating several skin disorders as it is significant to identify the features of disorder to distinguish between different types of lesions. The researchers have been ardently striving to develop an automated system to diagnose by being able to classify the images of several mild and malicious skin lesions. A deep learning based system [3] has been recently developed that has outperformed the general standard set by a number of dermatologists by passing the assessment and prediction criteria with a collection of many photographic and microscopic images of these skin lesions. Despite the fact that its implementation can be a bit computationally expensive yet the finalised model can be easily established on cellular telephones for global level improvement in the screening of such lesions. Fundus photography, i.e. a method used to take serial pictures of an eye via the pupil to record the appearance of an individual's retina in order to detect any abnormalities in it. It plays a major role in recognising the fatal conditions of the curable blindness [3]. Ophthalmologists use this technique to interpret while examining a large number of diabetic patients and it becomes difficult for them to work on it. Thus a solution has been proposed by a group of computer scientists and health specialists to recognize the diabetic retinopathy (DR), diabetic macular edema and other related eyesight problems via convolutional neural network by being able to use 128,175 images of retina [3]. These images have been trained by the model in order to do the work at ease and proved that deep learning can majorly contribute in extracting needed knowledge from the relevant raw data. This proposed system has been proven to be a success and was granted to be used by the healthcare providers due to exceptionally crossing the benchmark of pre-sensitivity and specificity set by the team to detect the causes and these diseases. Histopathology is a microscopic study of the affected tissues and cells with the aim to assist the histopathologists in treating the tissues and take care of the patients. It is in general an optimum way to diagnose various kinds of cancer [3]. These images featured from the study are hardly distinguishable by normal eyes but they can very well contribute in predicting possible results of the cancer patients. AI can devotedly stand in the way to detect prostrate and breast cancer with the advancement of deep convolutional neural networks. An automated system [3] is expected to be reducing the need of the pathologists by 2030 and provide an effective evaluation of the images gathered through histopathology as well as boost the quality care of these surviving patients. If the use of appropriate labelled data along with the deep-learning methods is applied onto such crucial medical fields then there is a higher possibility of bringing a thorough change in the healthcare practise in the system [3]. An automated Robotic surgery [3] that is absolutely managed by AI are widely utilised in the biomedical labs and assembly lines of assembly automation systems. Though the deployment of these robots for medical use have been taken considerably slow but the term synonymously goes well with the surgeries aided by the robots and still needs surgeons to operate an control the movements. AI has the capability [3] to improve medical-based decisions, enable diagnosis of the disease, recognizing prior unidentified images and patterns and aid in surgical actions for treating human diseases.

5   Cybersecurity in Healthcare

The present hike in breaching the cybersecurity of the healthcare system has kept the door open for the threat of exposing the private health data of the patients [10]. Along with this, several other risks are associated with it such as threats relating to financial and functional grounds of the healthcare system. The technological advancements have had several massive advantages in the healthcare field. It has been predicted that Electronic Health Records (EHR) [10] is about to mitigate the additional expenses of the healthcare by delivering enhanced quality services. Addressing these advantages [10], the Health Information Technology for Economic and Clinical Health (HITECH) Act has been approved while adopting Health Information technology (HIT) but this leaves a high increase in the probability of the health data breaches. Any available information on internet that has been lost, hacked or leaked to some other unauthorised recipients is referred to as a breach of cyber information and the interference is known as a cyber-attack [10]. It is assumed that the healthcare data is highly susceptible to cyber-attacks and financial threats associated with them as the data is high in volume. Private data records of patients [10] are more valuable to the hackers in compare to the data acquired from any other industries rather than the healthcare. The complications in providing cybersecurity surpasses the two major aspects as in, the financial loss of the health industry and privacy breach of the patient and includes one more risk to the patient's safety [10]. The cyber-criminals may use several techniques to alter the health data of the patient and expose him/her to life threatening situations. This situation can be extremely perilous for the patient and his/her health and thus, it is substantial to ensure that these health data is safe from exploitation. Several security methods [10] such as, encryption, use of strong passwords, etc. to ensure the limited disclosure of the health data to prevent cyber-attacks and mitigate the associated risks regarding the patient's health data and his/her safety. Despite the increasing risk of cyber-attacks in the healthcare system, there has been a huge literature gap in this specific research area [11]. There are various types of cybersecurity risks such as, Denial-of-Services Attack (DoS), Cryptographic Attack, Privilege Escalation, Malicious software, Trojan, Virus, Phishing, etc.. The people who ardently work on both securing as well as breaching the cybersecurity are: hackers, organizations, cybersecurity professionals, software developers, and end-users [12]. These vital cyber-security issues concerning threats to the global healthcare society needs to be resolved.

The policymakers are facing difficulties in filling up the gaps in bringing changes in the regulatory procedures as they have certainly attempted to develop policies for ensuring cybersecurity. With the regular advancement with time, the policies need to be updated as well so it becomes challenging for the policymakers to adapt instantly with the changing scenario. They should also make a note of the core programs arranged by the government which are facing severe cybersecurity threats as the telehealth service is widely adapted recently [10]. Thus, there is a dire need of dealing with the risks of cybersecurity faced by both society and government-based programs. Along with this, the involvement of blockchain technology [13] in such systems claims to provide highest security to the health data and thus to cope up to its aspects, it is needed that the policymakers keep on updating the regulatory environment according to the technological innovations. Healthcare systems face a number of challenges [10] that are still uncontrollable, but cybersecurity threats are those of all which in the long term has the capability to attain high success. A proper planning and implementation for ensuring cybersecurity can have a substantial impact on the healthcare society by strengthening the infrastructure.

6 Big Data in Healthcare

As health data is being produced at a massive amount and it comes from different point of resources, it is still debatable that how these data can bring revolutionary changes in the current healthcare system [14]. The term "Big Data" refers to a massive and diversified dataset which usually are not able to be conventionally analyzed and needs proper utilization to reach a long term goal in regard to bring the change in the societal perceived data. There are various medical data [14] with variety of states such as, some of them might be structured or unstructured data, either they can be research or patient oriented, and may be raw as well as explicit and implicit unprocessed data. Health data is very crucial and complex in nature that certainly needs to be secured than any other data in the industry. The entities that make the health data to be counted as the Big Data [14] are the number of participants involved in it as there may be situations where the community involved is global and needs to access the data from any point of location. Moreover, depth of composition should be maintained as to prepare and build a cohort it is significant to measure and choose the kind of information and understand the depth of it. Along with these, long term follow up is required that grants to observe the serial order of temporal events and so-on. There must remain a link between the sources of data so as to extract the required knowledge. Since these data are needed to answer a desired question based on any medical emergency, being able to compare and analyze is significant. Thus, the usage of data must be defined by a standard to be followed for maintaining harmony in the dataset [14]. The study [15] has proposed to introduce measures to protect the privacy of the health data as it has been witnessed that on the basis of cloud-based services the data are prone to several vital threats. The questionnaire served the purpose to highlight the drawbacks of the situation where these big data related to healthcare are cloud-oriented. It prominently enforces to adapt and pay more attention to the utilization of digital healthcare data and how to protect its privacy for generating decisions based on health data analysis [15].

7   Recent Attacks on Internet of Health Things

The modern technological advancements in healthcare system [16] involves wireless communication methods and medical devices along with wearable gadgets and body area networks (BANs). Internet of Things (IoT) majorly contributes in enhancing the overall patient care by allowing remote monitoring and other procedures related to the required treatment. However, the software and hardware requirements are complex in nature [16] and designing them on the basis of ensuring proper security and privacy is significant for the healthcare industry as violation in any of them can lead to severe disasters to patient's health conditions and providing the essential treatment on time. The current scenario of the attacks of Internet of Health Things (IoHT) portrays that the innumerous flaws in the design and the system lagging behind in ensuring security is paving way towards an insecure and unsafe healthcare system [16]. Since the subsisting modern technological healthcare infrastructure fails to meet the neccessary requirements for assuring privacy and security, the cyberattackers get immense opportunities to exploit different areas of the system. Several attacks on each of the components of the healthcare infrastructure have been discussed in order to give a clear picture on how these attackers are accomplishing targeted attacks on them [16]. Hardware attacks [16] in general refers to taking control of the weakness of the hardware device in order to misuse or perform malevolent activities remotely or physically. The attacker might have an open access to the internal structure of the hardware and might put in hardware trojans (HTs) to severely damage and corrupt the data leading to medical device failure. These trojans can be categorized into features based on action and physical attributes. There have been reports of several incidents [17] where the patient's medical data has been altered without any notice due to sudden insertion of HTs on the devices. An attack based on Hardware Trojans on the bacillus calmette guerin (BCG) was highly noticed as the attackers injected it to maliciously alter the logic behind the XOR gate in it. It is nearly impossible to identify such attack as it gets widely spread within the embedded system of the medical device [18]. The programs that are purposely written in a way to damage the healthcare system are referred to as Software attacks [18]. A number of risks concerning software and applications have been rapidly increasing with the increase in customised software and usage of embedded system in the healthcare sector. Malware software is the one created with an ill motive to have malicious impact on the community. Any medical device injected with malware can carry abnormalities in its casual operational functions. A malware named Conflicker attacked the Microsoft Windows Operating System in around 104 medical devices at a hospital in USA [19]. This granted the attacker to implement the code into the vulnerable system which made it impossible for the patients to reach and get their needed medical services from the hospital. Another widely used attack referred as Ransomware [18] that is a unique variant of malware used to restrict a user's access until a ransom is paid to unlock the system and relevant data. Around 50 hospitals of The U.K. were attacked by ransomware and this incident influenced and enforced the other healthcare organisations to close their computer system as a preventive measure [20]. This data breach had potentially disrupted the whole healthcare system by compromising the safety and privacy of the patient. Another common attack that is observed is the one that instantly focuses on the system. It reaches out to the system-level vulnerabilities via privilege escalation and weak authentication scheme and allows them to exploit them to obtain an unauthorised access to the health data [18]. The next one in the row of attacks is the side-channel attack that extracts sensitive health data by analyzing physical boundaries without intervention in the on-going process in the healthcare system [18]. Electromagnetic interference, spoofing of the sensors and the analysis of power attacks on health devices are the three categories of side-channel attacks that surpasses the physical proximity such as the structure of the circuit, etc. At present, wireless communication [18] is the preferred media to be used for remotely connecting the healthcare devices and attacks via these medium have become a common concern. The attackers can use denial-of-service attacks, battery depletion attacks, eavesdropping, etc. to harm these medical devices and interrupt the general platform for its regular operation.

The very recent attacks during pandemic COVID-19 that were held on the global platform of the healthcare organisations are summarized below: A report from the International Criminal Police Organization had been published in the year 2020 [21] to raise awareness due to the increase of attacks in the area of cyber world of healthcare during the pandemic of Corona Virus disease, namely, COVID-19. These recent attacks has been targeting all the public and private companies as well as every individual in the healthcare industry. The COVID-19 [21] has enforced the healthcare industry to be more technically influenced to include Information Technology into it and provide reliable medical care and necessary equipment to the patients. The industry has drastically taken the charge of relying on IT for storing information relating to vaccine, modeling the virus, etc. and this reliability is pushing the cyber-criminals to misuse the information and impart a negative impact on the crisis. In the last few months, the vulnerabilities have been globally misused such as the incidents where a cyber-attack stopped the absolute network of a hospital in Czech Republic, a ransomware attack was utilized on the month of March in the U.K. on a vaccine trial group, two unidentified attacks on the company building the hospitals chosen for being the centre for COVID-19 patients and on a Health agency in the U.K. and U.S.A. respectively (Table 1).

Furthermore, there have been severe attacks on the university research centres of UK, USA and Canada while developing vaccines in the month of July, 2020 to prevent COVID-19 disease which was supposedly a state-supported attack. Another report from the INTERPOL [21] claims that they have been receiving several threats by the cyber-criminals to halt the all around procedure to develop the vaccines otherwise will be held by ransom. The attackers are not limited only to the frontline services of the healthcare system during this pandemic [21] but also have been attacking the supply chain management of the medical industry on the same time. It seems that the prime targets of attack presently are the techniques developed for introducing vaccine and diagnosis or treating COVID-19 disease. Regrettably there are yet no certain ways to prevent the universities and hospitals from the cyber-attacks [21] and to fight against such severe issues thus leading to long term expenses and bad affects of these data breaches. The newest of Ransomware attack in the year 2018 on the UK's National Health Service had made them suffer a loss of around £92 million for both paying ransom and cancellation of the activities [21]. Therefore, it is substantial for the global healthcare system to be equally strong enough to pro-actively put effort in preventing and reducing the firm influence of such cyber-attacks.

Table 1 Reported cyber-attacks in healthcare system during pandemic

Date(s)	Place of cyber-attack
March 2020	Brno University Hospital, Czech Republic
March 2020	World Health Organization (WHO)
March 2020	COVID-19 Vaccine Trial Group, UK
March 2020	Paris Hospital, France
May 2020	Bam Construct and Interserve
June 2020	Babylon Health
July 2020	Authorities of the US, UK, and Canada

8 XAI for Trustworthy Internet of Health Things

There remains a dire need of transparency [22] in the healthcare system with the rapid increase of use of several types of structured and unstructured data along with the involvement of Artificial Intelligence for analysis of these data in the health community. The analysis of these health data requires a clean explanation when any certain outcome is predicted by models of the automated system. AI, unfortunately is incapable of providing this explanation and lags behind in ensuring transparency in the whole procedure. Here is where Explainable Artificial Intelligence (XAI) [22] is applied to enhance the circumstances created by the lackings of AI. The healthcare providers can rely on the overall picture generated by XAI as it boosts the trust between the system and the practitioners and researchers working in this field. It prominently enhances the chance of deploying AI in the healthcare sector as it has innumerous available techniques which can prove to be advantageous for the health community. It is significant to assure the proper utilization of the AI based healthcare system or health treatment model [22] to be accurate enough as it devotedly works towards saving human life. Furthermore, Smart Health Applications that involves wearable gadgets integrated with AI [2] are enabling the healthcare technology to experience a modification in the general infrastructure of the healthcare system. As these wearable gadgets are easily able to predict the user's or patient's health conditions via his/her health data analysis. XAI is highly recommended as a method that enables accountability, transparency, tracing of the possible outcomes and enhancing the models in the healthcare sector. Thus, to attain the confidence to treat a patient with an AI-based system XAI needs to be implemented in every running and upcoming model for ensuring to fill up the major gap in transparency and trust [22].

9 Methods for XAI Covering Cybersecurity and Healthcare

Machine Learning (ML) is being gradually preferred as the most convincing procedure to combat all the evolving risks of the cybersecurity community [22]. ML makes it easier for the users of the system to conveniently understand and interpret the operations of the system. The black-box models are used for making necessary decisions on the basis of predictions and it is certainly needed for the system to ensure to provide required trust and cleanliness to the users through the explanation given by it. Since the explanation supporting the binary outcome for the health data analysis is considered as the utmost desire of the community, many approaches have been focusing on the three aspects as follows: (a) The ML models used for the system must be able to provide a proper explanation for the result executed, (b) planned cyber-attacks on the interpreters within the set of white box, and (c) accurately providing definition of what and how much of the features and metrics of the interpretation should be generated. Yet the researchers have not been able to appropriately discover the issues and risks related to the cybersecurity sector and how these cyber-attacks on the black model in XAI. The study [23] proposed a classification of the procedures of XAI concerned with its relevant characteristics of security and risk models respect to the cybersecurity domain. They had designed a black box attack for the analysis of the basic components: consistency, correctness, and confidentiality of the XAI methods based on gradients. Their evaluation [23] shows how the method had easily succeeded in catching the attacker's malicious goal of confusing the classifier and the report of explanation generated by the model without changing the results.

10 Critical Analysis

We have reviewed few papers for the purpose of analyzing the state-of-the-art of the development concerning Explainable AI in the healthcare industry. There have been many proposals and attempts made on integrating AI/ML models with explanations in the healthcare sector. A novel technique has been proposed in the paper [24] which aimed to grant the AI system to work simultaneously with the specialists in taking decisions together. It uses decision tree as their approach for a moderate and aided clinical diagnosis. Another explainable model [25] for predicting multiple medications in a patient from their prescriptions of newly identified chronic disease was proposed. This work has the advantage of going hand-in-hand with the rate of increase in explain ability with its improved predictive performance. Zhenga et al. [26] had proposed an approach integrating deep learning to observe the motion and shape of the heart to categorize the cardiac pathology. Mimir [27] is an explanatory procedure that appends explanation to the Deep Neural Network (DNN) Model integrated in solving clinical issues by generating appropriate, structured and composed reports full of texts and images. Furthermore in another proposal given in [28], a procedure to understand the internal presentation of DNN models based on human-in-the-loop process for classifying mammograms by designating the behavioral patterns of internal units. Network Dissection (NetDissect) has been utilized for appraising explanation as a measure of how every unit of DNN sets with the human-understandable concepts. The short aforementioned analysis on how the XAI is changing the technological based healthcare industry portrays that there is a growing demand of integrating XAI into AI/ML based Healthcare projects so as to provide needed explanations and allows the specialists to understand the underlying logic of the predictions made by the machines. The system needs to be compatible and go with the flow of how human cognition level [1] works while making decisions technically.

11 Benchmark Datasets for Healthcare Cyber Security

As the cyber-attacks on the healthcare system have been rapidly growing, thus it has become significant to deploy strong preventive measures. Nevertheless, there is a shortage of available healthcare datasets due to privacy concerning the patients' data. An ECU-IoHT dataset [29] has been developed to contribute towards strengthening the cyber-security of the health ecosystem. It has been constructed in an IoHT ambience with several possible attacks to exploit the existing vulnerabilities of the set. At present, there is yet no other such available dataset on a public platform for assisting the community to analyze the behavioral patterns of attacks and evolving preventive measures for the same.

12 Future Challenges

Despite the fact that XAI [30] has been approached by the research community as a solution to the problems such as trust and reliance faced by AI based systems, there are still few challenges for XAI to overcome in Deep learning models within this while. There remains a lack of consent between the collection of vocabularies and several definitions concerning XAI. Again as the XAI is a new concept there remains a gap in having a benchmark terminology for the community. There is no option for trade-off within understanding and accuracy [30] and as XAI is especially introduced for generating convincing explanations, the motivation for an appropriate subjectivity to the terms can be taken from experiments on cognitive science. Another challenge for XAI methods involving Deep Learning is to address the accessible explanations and make them available for all the policy and lawmakers of the society.

13 Conclusions

This study has been focusing on Explainable Artificial Intelligence (XAI) which has been recognized as on of the utmost requirement of today's healthcare industry applications. XAI enhances the art of AI based healthcare systems as it allows to adopt the Machine Learning methods in real-life healthcare applications. The significance of XAI, both in general and in healthcare sector have been discussed here. The motive behind introducing the XAI in the healthcare industry: trust, transparency and accountability are the three basic components of healthcare systems which if included can highly change the nature of the healthcare services to the patients. The collected health data that is used to analyse the health conditions needs to be accurately trained by the ML models to improve the prediction accuracy in the systems. XAI is that particular area of expertise which focuses on providing the quality services to the users by interpreting the predictions made by the ML models in order to yield a global successful goal in the healthcare sector. XAI grants the AI to be more responsible and its vision to enhance the healthcare decisions taken by the system and strengthens its fundamental factors such as trust, transparency, accountability and privacy.

References

1. A. Adadi, M. Berrada, Peeking inside the black-box: a survey on explainable artificial intelligence (XAI). IEEE Access **6**, 52138–52160 (2018)
2. U. Pawar, D. O'Shea, S. Rea, R. O'Reilly, Explainable AI in healthcare, in *2020 International Conference on Cyber Situational Awareness, Data Analytics and Assessment (CyberSA)* (IEEE, 2020), pp. 1–2
3. S.R. Alotaibi et al., Applications of artificial intelligence and big data analytics in m-health: a healthcare system perspective. J. Healthc. Eng. **2020** (2020)
4. A. Holzinger, C. Biemann, C.S. Pattichis, D.B. Kell, What do we need to build explainable AI systems for the medical domain? (2017). arXiv preprint arXiv:1712.09923
5. M.A. Ahmad, C. Eckert, A. Teredesai, Interpretable machine learning in healthcare, in *Proceedings of the 2018 ACM International Conference on Bioinformatics, Computational Biology, and Health Informatics* (2018), pp. 559–560
6. N. Larburu, A. Artetxe, V. Escolar, A. Lozano, J. Kerexeta, Artificial intelligence to prevent mobile heart failure patients decompensation in real time: monitoring-based predictive model. Mob. Inform. Syst. **2018** (2018)
7. M.N. Burns, M. Begale, J. Duffecy, D. Gergle, C.J. Karr, E. Giangrande, D.C. Mohr, Harnessing context sensing to develop a mobile intervention for depression. J. Med. Internet Res. **13**(3), e55 (2011)
8. M.S. Hawley, S.P. Cunningham, P.D. Green, P. Enderby, R. Palmer, S. Sehgal, P. O'Neill, A voice-input voice-output communication aid for people with severe speech impairment. IEEE Trans. Neural Syst. Rehab. Eng. **21**(1), 23–31 (2012)
9. X. Du-Harpur, F. Watt, N. Luscombe, M. Lynch, What is AI? applications of artificial intelligence to dermatology. Br. J. Dermatol. **183**(3), 423–430 (2020)
10. S.S. Bhuyan, U.Y. Kabir, J.M. Escareno, K. Ector, S. Palakodeti, D. Wyant, S. Kumar, M. Levy, S. Kedia, D. Dasgupta et al., Transforming healthcare cybersecurity from reactive to proactive: current status and future recommendations. J. Med. Syst. **44**(5), 1–9 (2020)
11. M.S. Jalali, S. Razak, W. Gordon, E. Perakslis, S. Madnick, Health care and cybersecurity: bibliometric analysis of the literature. J. Med. Internet Res. **21**(2), e12644 (2019)
12. S.G. Langer, Cyber-security issues in healthcare information technology. J. Digit. Imaging **30**(1), 117–125 (2017)
13. M. Miliard, How does blockchain actually work for healthcare. Healthcare IT News April **13**, 2017 (2018)
14. S. Shilo, H. Rossman, E. Segal, Axes of a revolution: challenges and promises of big data in healthcare. Nat. Med. **26**(1), 29–38 (2020)
15. Z. Lv, L. Qiao, Analysis of healthcare big data. Futur. Gener. Comput. Syst. **109**, 103–110 (2020)
16. A. Newaz, A.K. Sikder, M.A. Rahman, A.S. Uluagac, A survey on security and privacy issues in modern healthcare systems: attacks and defenses (2020). arXiv preprint arXiv:2005.07359
17. M.E. Gladden, Information security concerns as a catalyst for the development of implantable cognitive neuroprostheses, in *9th Annual Conference of the EuroMed Academy of Business* (2016)
18. T. Wehbe, V.J. Mooney, O.T. Inan, D.C. Keezer, Securing medical devices against hardware Trojan attacks through analog-, digital-, and physiological-based signatures. J. Hardw. Syst. Secur. **2**(3), 251–265 (2018)
19. K. Fu, J. Blum, Controlling for cybersecurity risks of medical device software. Biomed. Instrum. Technol. **48**(s1), 38–41 (2014)
20. G. Martin, P. Martin, C. Hankin, A. Darzi, J. Kinross, Cybersecurity and healthcare: how safe are we? BMJ **358** (2017)
21. M. Muthuppalaniappan, K. Stevenson, Healthcare cyber-attacks and the covid-19 pandemic: an urgent threat to global health. Int. J. Qual. Health Care **33**(1), mzaa117 (2021)
22. D. Dave, H. Naik, S. Singhal, P. Patel, Explainable AI meets healthcare: a study on heart disease dataset (2020)

23. A. Kuppa, N.-A. Le-Khac, Black box attacks on explainable artificial intelligence (XAI) methods in cyber security, in *2020 International Joint Conference on Neural Networks (IJCNN)* (2020), pp. 1–8
24. I. Monteath, R. Sheh, Assisted and incremental medical diagnosis using explainable artificial intelligence, in *Proceedings of the 2nd Workshop on Explainable Artificial Intelligence* (2018), pp. 104–108
25. S. Kocbek, P. Kocbek, A. Stozer, T. Zupanic, T. Groza, G. Stiglic, Building interpretable models for polypharmacy prediction in older chronic patients based on drug prescription records. PeerJ **6**, e5765 (2018)
26. Q. Zheng, H. Delingette, N. Ayache, Explainable cardiac pathology classification on cine MRI with motion characterization by semisupervised learning of apparent flow. Med. Image Anal. **56**, 80–95 (2019)
27. S.A. Hicks, S. Eskeland, M. Lux, T. de Lange, K.R. Randel, M. Jeppsson, K. Pogorelov, P. Halvorsen, M. Riegler, Mimir: an automatic reporting and reasoning system for deep learning based analysis in the medical domain, in *Proceedings of the 9th ACM Multimedia Systems Conference* (2018), pp. 369–374
28. J. Wu, D. Peck, S. Hsieh, V. Dialani, C.D. Lehman, B. Zhou, V. Syrgkanis, L. Mackey, G. Patterson, Expert identification of visual primitives used by CNNs during mammogram classification, in *Medical Imaging 2018: Computer-Aided Diagnosis*, vol. 10575 (International Society for Optics and Photonics, 2018), p. 105752T
29. M. Ahmed, S. Byreddy, A. Nutakki, L. Sikos, P. Haskell-Dowland, ECU-IoHT: A dataset for analyzing cyberattacks in Internet of Health Things. Ad Hoc Netw. **122**, p. **102621** (2021). https://doi.org/10.1016/j.adhoc.2021.102621
30. A.B. Arrieta, N. Díaz-Rodríguez, J. Del Ser, A. Bennetot, S. Tabik, A. Barbado, S. García, S. Gil-López, D. Molina, R. Benjamins et al., Explainable artificial intelligence (XAI): concepts, taxonomies, opportunities and challenges toward responsible AI. Inform. Fus. **58**, 82–115 (2020)

Printed in the United States
by Baker & Taylor Publisher Services